GREEN RESEARCH, DEVELOPMENTS, AND PROGRAMS

GREENHOUSE GAS EMISSIONS AND NITROGEN LOSSES FROM GRAZED DAIRY AND ANIMAL HOUSING SYSTEMS

GREEN RESEARCH, DEVELOPMENTS, AND PROGRAMS

Additional books in this series can be found on Nova's website under the Series tab.

Additional e-books in this series can be found on Nova's website under the e-book tab.

GREEN RESEARCH, DEVELOPMENTS, AND PROGRAMS

GREENHOUSE GAS EMISSIONS AND NITROGEN LOSSES FROM GRAZED DAIRY AND ANIMAL HOUSING SYSTEMS

JIAFA LUO

AND

YAN LI

EDITORS

nova science publishers

New York

Library of Congress Cataloging-in-Publication Data

ISBN: 978-1-53611-100-2

Published by Nova Science Publishers, Inc. † New York

CONTENTS

PREFACE

Animal housing practices and stand-off pads (i.e., off-paddock or animal confinements) are increasingly being used by dairy farmers; largely to avoid crop and soil damage and for ease of herd management. This book assesses the impacts on greenhouse gas (GHG) emissions and nitrogen (N) losses from farming systems that include the use of animal confinements. This book consists of 10 chapters covering a number of aspects of animal "off-paddock" practices and associated gas emissions and N losses. It has been written so that each chapter can be understood separately without the need to read the whole book.

Chapter 1 describes the variety of off-paddock animal facilities employed on dairy farms in New Zealand and the extent of their use. The way these facilities are utilised and the resulting waste streams are discussed.

Chapters 2-4 deal with the gaseous emissions resulting from the collection and storage of effluents, slurries and manures and their application to land. The wide differences in GHG emissions from these systems due to differences in collection methods, amendments with litter or other materials, length of storage period, type of cover used during storage and method and timing of land application are discussed. A number of management options that can reduce GHG are identified and the need for a holistic view of management of emissions during all the stages involved in excreta management with consideration of the total GHG footprint is emphasised.

Chapters 5-6 focus on the use of dietary intervention or manipulation through incorporation of alternative forages or crops as an approach for increasing N use efficiency thereby either reducing the concentration of N in excreta, or changing the partitioning of N in urine and dung. The selection of forages for their potential to inhibit soil N transformations by either the exudation or urine-excretion of plant secondary metabolites is discussed. Additionally, an optimal cattle manure application rate to maximise crop yield and minimise risk of N loss to the environment is presented.

In Chapters 7 and 8 life cycle assessment (LCA) techniques applied to off-grazing technologies are reviewed and then applied to estimate GHG emissions from different types of housed/restricted-grazing systems used during winter in Southland and Waikato, New Zealand.

Chapters 9 and 10 present a review of the cost effectiveness of off-grazing technologies and an analysis of the costs associated with different types of housed/restricted-grazing systems used during winter in Southland and Waikato, New Zealand

From reading this book the readers will find that, as described in the executive summary, there is a requirement for future research, including the need for field measurements of methane (CH_4) and nitrous oxide (N_2O) emissions to quantify GHG emissions from each component of the "off-paddock systems" so that accurate data can be incorporated into total GHG footprint analysis and a fair comparison between all-year grazing and off-grazing systems can be made. The need to improve and validate both process-based and farm system models to help describe and understand GHG emissions from the complex systems involving off-paddock facilities is identified as well as the need to include a range of mitigation strategies.

The editors would like to express their appreciation to all the chapter contributors and peer reviewers for their dedicated effort and constructive comments. The editors acknowledge funding from the Taishan Scholars Program and the New Zealand Ministry for Primary Industries in the Sustainable Land Management and Climate Change programme.

Executive Summary

GAS EMISSIONS AND NITROGEN LOSSES ASSOCIATED WITH GRAZED DAIRY AND ANIMAL HOUSING SYSTEMS

Jiafa Luo*[*] *and Yan Li

AgResearch, Hamilton, New Zealand

Shandong Academy of Agricultural Sciences, Jinan, China

INTRODUCTION

Stand-off pads and animal housing practices (i.e., animal confinements) are increasingly being used; largely to avoid crop and soil damage and for ease of herd management. The use of animal houses and stand-off pads to avoid grazing during the risk periods for nitrogen (N) losses could be an effective greenhouse gas (GHG) mitigation technology and contribute to the overall goal to reduce GHG emissions from agriculture. However, there is uncertainty about unintended consequences (or 'pollution swapping') from these farm systems. Such consequences include potential nitrous oxide (N_2O), ammonia (NH_3) and methane (CH_4) losses from the animal confinement facilities themselves and following land application of effluent/manure collected from

[*]Corresponding author: Dr Jiafa Luo, AgResearch, 10 Bisley Road, Private Bag 3123, Hamilton 3214, New Zealand. Email: Jiafa.Luo@agresearch.co.nz.

these facilities. In this book impacts of the use of stand-off pads and animal housing on farm system GHG emissions and N losses are assessed.

Approach

A literature review and targeted field surveys were conducted to investigate animal confinements and their use in New Zealand dairy farming (Chapter 1). Data was collected from published literature and unpublished reports to assess GHG emissions, including N_2O, NH_3 and CH_4, from animal confinements and the collection, storage and subsequent land application of manure generated (Chapters 2-4). Some potential management practices and technologies for reducing GHG emissions from animal confinements and land application of animal manure and slurry were also evaluated (Chapters 2-6). System analyses and life cycle assessment (LCA) studies of restricted grazing technologies were reviewed to identify the carbon footprints and cost-effectiveness of whole-farm systems associated with animal confinements (Chapters 7 and 9). As well, LCA methodology and cost effectiveness analysis were used to assess the total GHG emissions of dairy farm systems involving off-paddock facilities over winter in Waikato and Southland in New Zealand (Chapters 8 and 10).

Off-paddock facilities (Chapter 1)

A variety of off-paddock facilities have been developed on dairy farms and can be grouped into four main categories: feed pads; stand-off pads; animal shelters; and wintering barns. Feed pads and/or stand-off pads are widespread on farms throughout New Zealand (~50% of dairy farms). While the use of animal shelters or wintering barns on farms is still very low (~2% of dairy farms), their use is gathering momentum as dairy production systems intensify.

These various off-paddock structures are used in different ways and for differing lengths of time. Generally the manure produced is not diluted with water. For stand-off pads the manure is combined with carbonaceous materials (e.g., wood chips, post peelings). As a result, various types of slurry and solid manures are produced that have different physical and chemical characteristics.

Gas emissions and nitrogen losses associated with grazed dairy and animal housing systems (Chapters 2-6)

When confined animals void their urine and dung on stand-off/feed pads or in houses, sheds or barns, it is generally stored for some time before subsequent application to land. There are wide variations in animal manure/slurry storage and management systems that include differences in manure collection methods, amendments with litter or other materials, length of the storage period, type of manure cover used during storage and method and timing of land application. Each step in the management of these manures/slurries has its own set of physical and environmental conditions affecting the production and emission of GHGs, thus each step must be managed independently. As a consequence, GHG emissions from these systems differ widely. It is important to highlight that management of these manures/slurries should be viewed as a continuum, where potential emissions from one stage are directly influenced by the conditions and consequent emissions that occurred in the previous stages. Therefore, reductions in GHG production at one stage may result in increased production at a later stage unless all stages are well managed. It is also important to note that there are several GHGs (N_2O, NH_3 and CH_4), and adjusting manure management practices to reduce one gas may increase the emission of another. Therefore, management of these emissions must be done with a holistic view of all the stages involved and with consideration of the total GHG footprint. There are a number of management options that can reduce GHG and N losses, including animal selection, feed choice and optimum application of slurry and manure to land.

Two case studies have compared GHG emissions from conventional grazing and off-paddock grazing systems. These studies compared N_2O emissions when stand-off pads were used to remove dairy cows during periods of potentially high N_2O emissions from grazed pastures. The use of the stand-off pad significantly reduced N_2O emissions (30-50%) from the grazed pastures. However, reductions from the entire farm systems were only 7% due to an increase in N_2O emission resulting from the land application of the collected manure/slurry. Due to the complexity and potential interaction between the factors affecting GHG emissions, both within and between management stages, systems based modelling should play an important role in designing manure management systems and assessing potential reductions in emissions when dairy cows are taken off the paddock.

Estimation of GHG emissions using LCA methodology (Chapters 7-8)

Gaseous losses from both animal confinement and pasture-based systems mainly occur as methane emission from enteric fermentation. Other hotspots in the confinement systems are methane and N_2O emissions from manure storage and management and GHG emissions associated with the production of food concentrates (e.g., grains). The limited number of international LCA studies, which compared pasture-based and confinement-based dairy systems, were hard to compare as a result of differences in methodologies used. However, in most studies the confinement-based systems had a lower methane emission from enteric fermentation resulting from digesting less pasture and more concentrates and/or alternative feed supplements, while GHG emissions (mostly methane and N_2O) associated with manure storage were higher in these confinement-based systems.

An LCA methodology was used to assess the GHG emissions of several restricted grazing strategies involving use of stand-off pads or housing in the Waikato and Southland regions of New Zealand. Results were highly dependent on the CH_4 and N_2O emission factors used for the analysis. This model analysis indicated that the use of stand-off pads or housing for wintering dairy cows can lead to an increase in the carbon footprint of milk produced from dairy farm systems, where there are no other farm system changes. This is due in part to increased use of fossil fuels with cut-and-carry of pasture and establishment and maintenance of infrastructure. Although several studies found lower total N_2O emissions from the housed system over winter compared to all-year pasture, this LCA analysis suggested that the main increase in GHG emissions was due to management of manure/slurry collected from the pad or housing systems and particularly due to methane emissions from stored manure/slurry. If feeds with a relatively high carbon footprint, such as palm kernel expeller (PKE), are used during housing, there can be an even larger increase in the carbon footprint of milk produced from dairy farm systems. Manure/slurry storage length and conditions (such as bedding materials in animal housing) can have significant impacts on the carbon footprint. For example, amending manure/slurry with sawdust or straw significantly reduces the carbon footprint if stored for 3 months compared to 7 months.

Cost-effectiveness analysis of off-grazing technology (Chapters 9-10)

Cost-effectiveness analysis compares the relative costs and outcomes (or effects) of alternative courses of action. The most significant costs associated with off-grazing technologies are the building (or capital cost), and costs associated with feed and manure/slurry handling, treatment and application. Costs that vary with operation level (e.g., herd size) include labour costs and manure management. From overseas information, use of animal housing involves high capital costs. These are often managed through economies of scale as a result of reduced cost per head associated with larger herd size and hence larger housing structure. Level of confinement, and feed and manure/slurry management practices can also be varied to reduce the overall cost of animal housing while also increasing the efficient use of farm inputs such as fertilizer, labour and feed.

Cost effectiveness analysis was used to assess the cost per tonne of CO_2-e GHG emissions from the different wintering systems. The results give an indication of how the systems compare in terms of costs and reduction in GHG emissions. The capital cost of housing is the main component of the costs associated with wintering strategies. Stand-off pads are cheaper than deep litter barns which are cheaper than slatted floor barns. Other important costs are operational costs associated with animal feed and manure and slurry management. The cost saving from reduced N application is slightly greater than the additional cost of labour required for cutting and carrying pasture to cows in housing. However, this cost saving is less than the additional cost of off-farm pasture silage supplements. Generally, wintering on pasture is more cost effective per unit of GHG emission due to lower fixed and operating costs per cow and lower GHG emissions per cow.

Future research

Due to the increased use of off-paddock systems, there is a need for a more comprehensive understanding of interactions of GHG emissions within and between each step of these systems. Reductions in GHG production at one stage of the manure/slurry management system may result in increased production and emissions at other stages, while mitigation of one gas may lead to increased emissions of another or affect nutrient loss after application of manure/slurry to agricultural land. To help improve understanding of these processes, and identify effective mitigation strategies at the farm system level, the following future research activities are recommended:

- This review suggests that, using current limited data, emissions of methane and N_2O are dominant contributors to the higher carbon footprint from stand-off/housing systems. Quantification of GHG emissions from each component of the "off-paddock systems" by conducting field measurements of methane and N_2O emissions is required, so that accurate data can be incorporated into the total GHG footprint analysis and a fair comparison between all-year grazing and off-grazing systems can be made.
- Validation and improvement of both process-based and farm system models are needed to help describe and understand GHG emissions from the complex systems involving off-paddock facilities. As well, more comprehensive farm system modelling is required to optimise farm systems where stand-off pads or animal housing are used. This could also include optimal use of different types of supplementary feed. These aspects would also need to be modelled using an LCA approach to ensure that all system related emissions (including production and use of feeds) are included in the analysis.
- Further model evaluation, including the use of LCA, of optimised systems with a range of mitigations included is required. Examples of mitigations could include integration of a system of methane capture and energy use, and use of low carbon footprint feed sources or alternative forages. These mitigation strategies need to be tested and validated with on-farm systems trials.

In: Greenhouse Gas Emissions ... ISBN: 978-1-53611-100-2
Editors: Jiafa Luo and Yan Li

Chapter 1

OFF-GRAZING ANIMAL CONFINEMENT FACILITIES AND THEIR USE IN NEW ZEALAND

Bob Longhurst*[*] *and Jiafa Luo
AgResearch, Hamilton, New Zealand

1.1. INTRODUCTION

New Zealand dairying has traditionally been a pasture based grazing system with conserved feeds, hay and silage, made in times of pasture surplus (spring) and fed to cows in the winter months when pasture growth is slow. Farmers stocked their farms with ~2.5 cows/ha (DairyNZ production system 1) and used rotational grazing of paddocks to achieve and maximize pasture quality (DairyNZ 2014a). Usually a 'day and night' paddock approach was employed and livestock were rarely removed from pastures when wet. Later, the concept of standing cows off on farm races was used but farmers soon realised that this practice quickly caused massive damage to races that required expensive remediation. Additionally, regional councils became increasingly concerned about uncontrolled runoff from this practice contaminating waterways. The use of 'sacrifice paddocks' was another wintering practice used by farmers to protect pastures. Farmers would choose

[*]Corresponding author: Mr Bob Longhurst, AgResearch, 10 Bisley Road, Private Bag 3123, Hamilton 3240, New Zealand. E-mail: Bob.Longhurst@agresearch.co.nz.

a paddock planned for pasture renewal and contain stock in the one paddock for as long as possible. While sacrifice paddocks are not in themselves illegal, any discharge of effluent or sediment into ground or surface water is prohibited. As a result, sacrifice paddocks are a high-risk wintering option when it comes to protecting water quality and soil structure. Grazing on wet soils can lead to degradation of soil structure and can affect pasture quality and subsequent production (Ledgard et al. 1996; Singleton et al. 2000). These research findings on the soil damage caused by stock treading and its impact on limiting pasture production have made many farmers aware of the need to remove cows from susceptible pastures, especially during wet periods.

All New Zealand dairy farms are categorised into one of five production systems based on the timing, purpose and amount of imported feed use, both purchased as supplements and grazing off for dry cows (DairyNZ 2015a). Grazing policies of young stock are excluded when categorising herds into the five production systems described below.

- **System 1:** All grass, self-contained – No supplementary feed purchased. No cows grazing off the milking area.
- **System 2:** Dry cow feed purchased – 4-14% of total feed imported and fed to dry cows including dry cows grazing off the milking area.
- **System 3:** Feed purchased for dry cows and to extend lactation – 10-20% of total feed is imported to the milking area (usually autumn feed) and for dry cows.
- **System 4:** Feed purchased for dry cows and to extend lactation – 20-30% of total feed purchased at both ends of lactation.
- **System 5:** Feed purchased for all year round feeding – at least 30% of total feed is imported including for dry cows.

In this publication, we make reference to Low, Medium and High input systems which DairyNZ have defined (DairyNZ 2015a) as:

- **Low:** System 1 and System 2.
- **Medium:** System 3.
- **High:** System 4 and System 5.

Many New Zealand dairy farmers have moved away from Low production system farms over the last 10-15 years, adopting higher input systems as they try to manage volatility in milk prices and feed input costs. This trend is highlighted in Table 1.1 and shows that over the past 13 years Low system farms have decreased from 70% to 30% of all dairy farms (Mounsey 2015). Conversely, the numbers of Medium system farms have increased during the same period from 17% to 41% of all dairy farms while the proportion of intensive High input system farms has increased from 13% to 29%.

Table 1.1. Trends in New Zealand dairy farming

Dairy Statistics	2000/01	2013/14
Farm data[1]		
Dairy herds (no.)	13,892	11,927
Average herd size (no. cows)	251	413
Average farm size (ha)	96	144
Average stocking rate (cows/ha)	2.62	2.87
Dairying trends		
DairyNZ farm production system [2]		
Low (%)	70	30
Medium (%)	17	41
High (%)	13	29

Source: [1]DairyNZ (2014a); [2] Mounsey (2015).

The use of stand-off pads and animal housing (i.e., animal confinements) is increasing on New Zealand dairy farms because of this recent intensification. Their use is also promoted to reduce nitrogen (N) losses from grazed pastures and it is often claimed that both animal welfare and productivity are improved. Housing of animals may reduce the emission intensity of enteric methane (CH_4), as housed animals could be expected to have lower maintenance requirements since energy costs for walking and grazing activities may be lower. However, alongside potential benefits, there is a risk of unintended consequences (or 'pollution swapping') from these systems. Such consequences include potential nitrous oxide (N_2O), ammonia (NH_3) and CH_4 losses from the confinements themselves and following land application of effluent or manure collected from these confinements (Chapters 2-4).

1.2. Types of Off-Paddock Facilities

A variety of off-paddock facilities have been developed on dairy farms as a result of the rapid intensification in the industry over the past 10-15 years.

Off-paddock options for wintering stock used by dairy farmers were identified by Longhurst and Lambourne (2006).

The most common types of off-paddock facilities can be grouped into four main categories:

1) Feed pads: used for supplying supplementary feeds to stock.
2) Stand-off pads: used to protect soils.
3) Animal shelters: multi-purpose use and mainly short-term animal confinement.
4) Freestall barns: generally longer-term animal confinement.

However, there are variations on each of these main categories and each is discussed in detail in the following sections.

1.2.1. Feed Pads

Structure

Feed pads are hard surface platforms (usually concrete) sited handy to the milking shed where stock can be fed supplementary feeds (Figure 1.1). Many farms now import supplementary feeds to complement the pasture diet and using a feed pad achieves greater feed utilisation than if fed out in a paddock. As dairy herds generally come to the milking shed twice daily during lactation, extra rations (e.g., maize silage, palm kernel expeller (PKE)) or mineral additives (e.g., lime flour, salt, or magnesium) can be more easily introduced and mixed into the cows' diet when fed on the feed pad.

Figure 1.1. Cows on feed pad.

Stock duration on the feed pad is normally limited to 1-2 hours per session around the morning and afternoon milking's. Therefore, the typical time spent on a feed pad would be around 3 hours per day. However, as the trend in dairying moves to more intensive systems (DairyNZ systems 3, 4 and 5) with greater amounts of feed supplementation being fed to cows, the time spent on feed pads is likely to increase to 4-6 hours per day. An area of 3.5 m^2 per cow is required when cows are using feed pads for short periods of time (DairyNZ 2014b).

Effluent Management

Effluent management is normally carried out by scraping the pad surface using tractor-mounted rubber scrapers to remove excreta (Figure 1.2). A minority of farms employ flood washing systems for cleaning the pads. Depending on the slope of the feed pad, excreta can be separated into liquid and solid components or simply scraped as slurry into a holding tank or storage pond for further treatment along with the farm dairy effluent. As regional councils now require effluent to be contained, most farms scrape solids to a dedicated concrete storage bunker. The bunkers are normally uncovered and solids are stored for 3-6 months before land application.

Figure 1.2. Pad manure scraped to sump.

Feed pad variant

One variation used for feed pads is to have the pad covered (Figure 1.3). This practice usually doubles the cost of construction but can be valuable in areas where limiting the increase in volume of effluent due to rainfall is important and savings can be made on the size of the effluent storage pond. The volumes of effluent produced are therefore limited to the amount of excreta deposited by stock.

Figure 1.3. Covered feed pads.

1.2.2. Stand-Off Pads

Structures

Stand-off pads (also referred to as loafing pads) are specially built uncovered areas where cows can be withheld from grazing pastures during wet periods to rest. This practice minimises damage to soil structure and consequently can improve pasture production. Another effect, with major greenhouse gas (GHG) implications, is that the proportion of excreta, particularly urinary N, deposited onto pasture is reduced. As a consequence, both nitrate (NO_3^-) leaching and N_2O emissions during wet periods can be reduced.

The stand-off pads are normally constructed of a free-draining material on which cows can lie (Figure 1.4). Carbonaceous materials, such as wood chips, post peelings, bark or sawdust are commonly used as bedding. However, other inert materials such as lime or soft metal (rock) mix, even river stones, are sometimes used if the cost of obtaining carbonaceous materials is prohibitive. The longer the cows are held on a stand-off pad, the larger the area required per cow (Table 1.2). For lactating cows held short term on a stand-off pad an area of 1 m^2/100 kg liveweight is recommended (DairyNZ 2014b).

Table 1.2. Minimum area required per cow on woodchip for different stand-off situations

Time on stand-off pad	Area per cow
Short term (12 hrs/d for up to 2 days in a row)	$3.5m^2$
Long term (>12 hrs/d for 3 or more days in a row)	$5.0m^2$
Permanently (self-feeding system including feeding area)	$9.0m^2$

Source: DairyNZ (2014b).

For a stand-off pad to function properly it is essential to have a good drainage system in place. The normal practice is to have a series of sub-surface drainage coils (usually a network of Novaflow piping) (Figure 1.5), overlaid with drainage metal followed by carbonaceous materials (Figure 1.6). The pads are constructed so that drainage falls to a central low point from where it is piped to an effluent sump or storage pond.

The use of stand-off pads is more common on farms with higher-risk (poorer draining) soils or in areas subject to higher rainfalls (>1,400mm per year). Their usage is popular in Northland, Waikato, Taranaki, Westland, Otago and Southland. Pad usage on farms is normally confined to between drying cows off (May) and calving (July-August) and usually covers 8 weeks (range 6-12 weeks). Typical stock usage is to hold cows overnight for 16 hours per day. However, there is generally a period during each winter when climatic conditions are sufficiently bad for stock to be held on the pad for the entire day.

Figure 1.4. Post peelings bedding on a stand-off pad.

Figure 1.5. Drainage coil overlaid with metal (photo credit: Wendy Sheppard).

Figure 1.6. Drainage metal overlaid with a layer of 30-40 cm wood chunks before final bedding layer of 50 cm wood chips or post peelings (photo credit: Wendy Sheppard).

Effluent Management

Leachate from stand-off pads is collected into a mainline drain and piped to a storage pond for treatment in the normal farm dairy effluent management system. The draining leachate can have high concentrations of nutrients, particularly N and potassium (K). These concentrations can be several times the concentrations found in typical farm dairy effluent.

A surface layer of the solid material (containing dung) on the pad is usually scraped off the surface and stockpiled for 2-3 months until it is required as a soil conditioning amendment for land destined to be cropped. From the time when the carbonaceous material is first placed on the pad surface until the time it is removed, changes in its physical, chemical and microbial composition take place. These changes include rapid nutrient enrichment, decreased C/N ratio and increased alkalinity (pH 7.5-8.0) (Luo et al. 2008).

Stand-Off Pad Variant

A variation on the stand-off pad is the self-feeding pad. The self-feeding pad normally has a concrete platform for a silage stack along the length of one side. Throughout the wintering period the silage stack is progressively fed out to the stock. This arrangement is popular in the Otago and Southland regions for total wintering of cows over 10-12 weeks. Because cows are held for 24 hours per day and seven days per week, the area required per cow is larger than that required for stand-off pads (9 m^2/cow, i.e., 8 m^2 for lying area and 1 m^2 for feeding area) (DairyNZ 2014b).

Common pad management is to scrape manure from the concrete feeding apron in front of the silage stack and remove it to an adjacent concrete storage bunker where the manure could be held for 3-4 months until applied to land in late spring. The concrete feeding pad is usually constructed with a slight fall so that liquid effluent can drain to a storage pond in the opposite direction to where the solids are scraped.

1.2.3. Animal Shelters

The use of animal shelters (or loose housed systems) has been steadily gaining in popularity since year 2000. These structures perform the function of a feed-pad, stand-off pad and covered shelter. There are two main systems of animal shelter used in New Zealand dairying: HerdHomes® and Redpath®.

Herdhomes® Shelter

Structure

The HerdHomes® design includes a clear plastic roof and a concrete feeding strip along each length of the structure (Figure 1.7). The "floor" is constructed from specially designed concrete slats that allow cows to stand for prolonged periods without damage to hoof and legs. Below the slatted floor there is a 1.2 m deep concrete lined manure bunker designed to store excreta.

Figure 1.7. HerdHomes® animal shelter.

Figure 1.8. Concrete slatted floor in HerdHomes® shelter.

Stock usage of the HerdHomes® shelter varies by region both in hours used and in effluent management approaches (Table 1.3). Most farmers use the HerdHomes® shelters intensively throughout the wintering period and into spring if adverse weather conditions are encountered. For the rest of the year the shelters are used either on an 'as needed' basis or daily as a feed pad. In Northland and Waikato in the northern part of New Zealand, the shelters are often used with a shade cloth (Figure 1.8) to alleviate heat stress for cows (4-6 hours per day) over humid summers when temperatures reach the high 20s Celsius (Pow et al. 2010).

Stocking densities in the shelters vary depending on the cows' breed and mob (dry, springer, milking, or colostrum) at a particular time of year. However, the basic guideline is to provide 1 m^2 of space in the shelter for each 100 kg of cow liveweight (Pow et al. 2010). For short-term use (<8 hours/day) a common stocking density is about 200 cows/HerdHomes® shelter (3 m^2/cow).

The HerdHomes® shelters are used extensively as feed pads to complement the herd's diet with silage, grains, process by-products and minerals. Southern South Island farmers have reported a lower feed requirement for wintering in a shelter compared to on a crop; 8 vs 14 kg DM/cow/day (Longhurst et al. 2006).

Table 1.3. Typical characteristics of HerdHomes® shelter usage by most farmers

	Northland	Waikato	Westland	Southland
Stock usage (hrs/day)[1]				
Winter (months)	20-24 (2)	6-14 (2)	12-22 (3)	24 (3)
Spring	4-12	2-12	4-8	3-12
Summer	2-8	5-6	1-2	1-2
Autumn	0-10	4-6	1-2	1-5
Effluent management				
Effluent type	Solids	Slurry	Solids	Slurry
Liquid drained	Yes	No	Yes	No
Times emptied/yr	1-2	1-2	1-3	2
Applied to land	Spr-Aut	Spr-Aut	Aut	Aut-Sum
Appl. Rate (t/ha)	10-30[2]	10-30[2]	10	10

[1]Zoe Pow, HerdHomes® Ltd, pers. comm.

[2]Higher application rate when applied to land for maize crops.

Effluent Management

The under-floor manure bunkers of HerdHomes® shelters have the capacity for holding up to 500 m^3 of manure (Figure 1.9). This capacity means that the bunkers can usually hold at least 3-months manure storage with continuous use. A deeper manure bunker (1.5 m) is an option that southern South Island farmers use if their stock totally wintered for up to 12 weeks.

Figure 1.9. Underfloor manure bunker.

There are a large number of HerdHomes® structures in Northland as they were the invention of Northland dairy farmers Tom and Kathy Pow. However, their popularity has spread throughout the regions, especially Southland, Otago and Westland.

Redpath® Shelter

Structure

The Redpath® shelters are clear plastic roof loose housing structures covering a floor of carbonaceous material (Figure 1.10). Usually at least 30 cm of wood chips or similar carbonaceous material provides the bedding for resting stock. Stock can be fed via feeding lanes that can be either internal or external to the structure.

Farmers use these shelters to winter dry cows for 12-16 hours per day, and even 24 hours during severe weather conditions. Calving is also likely to occur in these shelters. Once calved, the colostrum mob is likely to use the shelters for 12 hours per day for 5-6 days while the milking cows are likely to use the shelter as a feed pad for 2-3 hours per day throughout the early spring period.

Figure 1.10. Redpath® shelter with internal feeding lanes.

Effluent Management

Drainage under the carbon bedding normally comprises shallow humps with Novaflow piping in the hollows, topped with pea metal. Some farmers use a "mud cloth" to separate the metal and carbon layers. Drainage appears to be small from these covered shelters (~2 L/cow/day). A standard practice is to collect any leachate into an underground tank for later land application to pasture.

The carbonaceous bedding material is normally removed after the wintering period, stockpiled for up to 3 months and then spread over an area designated for cropping during the months of October and November in the spring.

Table 1.4. Typical carbon and nitrogen concentrations (on "wet weight" basis) in uncovered and covered stand-off pad media

Stand-off structure	%DM	%N	%C	C/N ratio	pH
Uncovered	29	0.21	13.7	65	7.7
Covered	33	0.30	11.9	40	7.9

Source: unpublished data (Bob Longhurst).

Because the carbonaceous media is covered there is no dilution or leaching of nutrients by rainfall. As a result the nutrient content in the carbonaceous materials is higher compared to uncovered stand-off pad material (Table 1.4).

1.2.4. Freestall Barn

Structure

The introduction of totally enclosed housing of dairy cows is relatively new in New Zealand. The freestall barn designs tend to be built on the Europeans' experience with housing animals indoor during winter. Freestall barns are used by some DairyNZ High system farmers (Table 1.1) to extend lactation to 305 days per year, maximise feeding potential and to minimise environmental impact.

Freestall barns are purpose designed steel and iron buildings that incorporate feeding lanes, resting cubicles and walking/excreta lanes (Figure 1.11). They generally use concrete floors. However, in the cubicles where cows lie, farmers can choose several bedding options: foam or rubber mattresses, waterbeds or sand (DairyNZ 2015b).

Figure 1.11. Freestall barn showing roofing designs, cubicles and manure lanes.

Once under cover, cows are free to rest, roam or feed at their leisure. There should be no dead ends in the building with cow traffic and access to feed and water a design priority. Ultimately the feed space per cow drives the length of the building e.g., 70 cm per cow with 100 stalls on each side will equate to a 70-metre barn in length (70 cm per cow x 100 cows = 70 m). The total width of the barn is driven by the number of rows of stalls and the width of the passageways (DairyNZ 2015b).

As with animal shelters, farmers have noted that the use of freestall barns reduces feeding requirements. One Southland farmer calculated cow energy requirements reduced from 200 MJ ME/day when stock were grazing brassica crops to 96 MJ ME/day when wintered indoors. Southland farmers typically have a dairy support block close to the milking platform (<10 km) where pasture silage and whole-crop silage are grown for winter feed. The fertility needs of these crops are in turn supplied from the application of freestall barn manure.

Effluent Management

These structures are designed to capture effluent in the walking lane which normally employs an automatic cleaning system using hydraulic scrapers to remove excreta to a central drain, typically every 3 hours during wintering (Figure 1.12). Outside of the wintering period manure scrapings would depend on intensity of barn use, but if used as a feed pad for 2 hours then one scraping per day would suffice.

Figure 1.12. Automatic scraper in action.

A freestall barn example from Southland was studied by Houlbrooke et al. (2011). They found that the scraped excreta had the following chemical characteristics: DM 8%; total N 0.32% (wet base); organic C 3.2% (wet base). Manure with these characteristics is described as slurry. Over 70 days, 21,000 m^3 of slurry is produced from 600 cows and held in a storage pond where it is diluted as a result of being exposed to natural rainfall. The pond slurry is normally applied to pastures via a 14 m^3 slurry tanker at rates of 12-20 m^3/ha (Houlbrooke et al. 2011). The latest land application techniques, such as soil injection, are being employed on some farms to minimise gaseous losses. A large tractor is required for soil injection of the slurry.

An alternative approach to effluent management is to have the manure scraped to a large sump and from there pass through a mechanical solids separator. The solids are then stored on a concrete pad, possibly covered from rain, and then spread using a muck spreader at the earliest opportunity. On many farms this means that solid manure applications are occurring during winter as there is not sufficient solids bunker space for storage (usually about 2 weeks only) over an entire wintering period. Post-separation liquid is stored in a large effluent pond and applied to land via a low rate effluent irrigation system.

Interest in housing cows and high input systems is strong in Otago and Southland in southern New Zealand due to a combination of factors:

- above average size herds (>550 cows), making it more economical.
- soils susceptible to pugging.
- harsher winter climatic conditions.

This interest is likely to increase in the future, particularly if the milk solids price rises above $7/kg MS (Megaw and McCall 2012).

1.3. Stock and Animal Health Issues with Off-Paddock Facilities

1.3.1. Animal Health

Stock metabolic problems, such as milk fever or grass staggers, could be thought to be more common under animal confinement conditions. However, farmers in southern New Zealand report that since using these structures they have better control through their stock feeding system and the input of the right minerals (e.g., magnesium oxide, lime flour) to the cows' diet. As a result, there can be fewer metabolic cases compared to traditional farming practices. If cows are well fed and an optimum pre-calving diet is given, there is less risk of Fatty Liver Syndrome. Farmers report that stock abortions are lower with housing compared to stock being wintered on brassica crops. Farmers monitoring cow blood status have found higher copper levels in housed animals compared to those on pasture/crops.

Farmers report that animal health is generally better with housing systems. Lameness can be reduced as cows are not subjected to walking on wet tracks in winter/spring. Mastitis levels appear unaffected, but farmers report that extra work is required to ensure that the cubicle beds and concrete areas are kept clean. Animal health of confined stock must comply with welfare standards (MPI 2014).

1.3.2. Young Stock

Calves tend to leave the farms at weaning to graze off-farm and return as in-calf rising 2 year old heifers. This would normally occur in May with the heifers introduced to the barn and wintered in the barn until calving in late July. Some freestall barns have a dedicated calving bay with straw laid over the concrete for cow comfort.

1.4. Numbers of Off-Grazing Facilities

As there is very little published data, except for MAF (2008) and WRC (2012), on the types and numbers of off-grazing facilities, this information was mainly sourced directly from dairy companies (particularly Fonterra Ltd), companies involved in building animal confinement structures, regional councils and dairy industry personnel. This information was gathered through correspondence, telephone calls and direct contact with the relevant parties. Where firm data was not available, the best estimate was sought from various people. Information from these sources was then collated into a database to provide an overview of the present status of these off-paddock facilities in New Zealand (Table 1.5).

Table 1.5. Dairy statistics for 2010/11 milking season and estimated number of off-paddock structures on New Zealand dairy farms

	North Island	South Island	New Zealand
Dairy statistics			
Number farms	8,947	2,788	11,735
Average farm area (effective ha)	119	205	140
Average herd size (no. cows)	325	582	386
Milk production			
Milk solids (kg MS/ha)	866	1,030	923
Milk solids (kg MS/cow)	319	362	334
Structures used (% of all dairy farms)			
Feed pads (%)	28	25	27
Stand-off pads (%)	24	15	22
Animal shelters (%)	1.5	2.4	1.6
Freestall barns (%)	0	1.8	0.5

In 2008, veterinarians conducted a nationwide survey of their dairy farming clients in an attempt to get an overview of the types of housing systems currently used and the management practices associated with their use (MAF 2008). Findings from a combination of veterinary surveys (956 farms), phone interviews with producers (30) and regional workshops (2) confirmed that no more than 3% of dairy farmers were using some form of animal housing. A 2011 survey of 401 Waikato farmers by the WRC (2012) found that 24 and 22% of farms had feed pads and stand-off pads, respectively.

While the use of stand-off pads and feed pads is now widespread throughout New Zealand, the use of animal shelters or freestall barns is still relatively low, although gathering momentum. Some of the major potential benefits of dairy housing identified by Longhurst et al. (2007) were pasture protection, effluent management, cow health and welfare, supplementary feeding management and milk production benefits.

CONCLUSION

New Zealand dairying traditional all-pasture based grazing systems are rapidly changing. Farmers desiring to protect soils from stock treading damage and to achieve higher feed conversion efficiencies are increasingly using off-paddock structures. These structures include feed pads, stand-off pads, animal shelters and freestall barns. Feed pads and/or stand-off pads are widespread on farms throughout New Zealand (~50% of all farms). While the use of animal shelters or freestall barns on farms is still relatively low (~2% of all farms), their use is gathering momentum as dairy production systems intensify.

Each of these off-paddock structures are used in different ways and for differing lengths of time. Generally the manure produced is not diluted with water. For stand-off pads the manure is combined with carbonaceous materials. As a result various types of slurry or solid manures are produced that have different physical and chemical characteristics. Gaseous emissions will occur at deposition, during storage and at time of land application and these losses are further explored in the following chapters (Chapters 2-4).

Acknowledgments

The authors acknowledge funding from the Ministry for Primary Industries of New Zealand (MPI) through the Sustainable Land Management and Climate Change (SLMACC) programme.

References

DairyNZ (2014a) *New Zealand Dairy Statistics 2013-2014.* DairyNZ Limited, Hamilton, New Zealand.

DairyNZ (2014b) *Stand-off pads. Your essential guide to planning, design and management.* DairyNZ Limited, Hamilton, New Zealand.

DairyNZ (2015a) *DairyNZ Economic Survey 2013-14.* DairyNZ Limited, Hamilton, New Zealand.

DairyNZ (2015b) *Dairy cow housing. A good practice guide for dairy housing in New Zealand.* DairyNZ Limited, Hamilton, New Zealand.

Houlbrooke D., Longhurst B., Orchiston T. and Muirhead R. (2011) *Characterizing dairy manures and slurries.* Envirolink tools report AGRX0901.

Ledgard S. F., Thom E. R., Singleton P. L. and Edmeades D.C. (1996) Environmental impacts of dairy systems. *Proceedings of Ruakura Farmers Conference* 48: 26-33.

Longhurst B. and Lambourne A. (2006). What options are there for wintering cows off pasture? *Dairy3 Conference Proceedings* 4: 3-10.

Longhurst B., Binnie B., McDermott A. and Oliver L. (2007) *Benefits of housing or partially housing NZ dairy cows*. Dexcel and Dairy InSight Report.

Longhurst R. D., Luo J., O'Connor M. B. and Pow T. (2006) Herd Homes: a new innovation in dairying to help achieve the 4% per annum production goals. *Proceedings of the New Zealand Grassland Association* 68: 309-313.

Luo J., Ledgard S. and Lindsey S. (2008). A test of a winter farm management option for mitigating nitrous oxide emissions from a dairy farm. *Soil Use and Management* 24: 121-130.

MAF (2008). *Impacts of New Zealand housing systems on dairy cow welfare*. MAF Biosecurity New Zealand Technical Paper No.: 2009/25.

Megaw A. and McCall L. (2012) *Dairying in Southland. Opportunities of changing land use*. New Zealand Grassland Association On-farm field tour programme.

MPI (2014) *Dairy Cattle – A Code of Welfare issued under the Animal Welfare Act 1999*. Ministry of Primary Industries, New Zealand Government, Wellington, New Zealand.

Mounsey Z. (2015) *Analysis of Production Systems in the New Zealand Dairy Industry*. Report for Kellogg Organisation, pp 42.

Pow T., Longhurst B. and Monaghan R. (2010) *Best practice manual for HerdHomes® Systems: Guidelines for owners and operators*. HerdHomes Systems® Ltd, Whangarei.

Singleton P. L., Boyes M. and Addison B. (2000) Effect of treading by dairy cattle on topsoil physical conditions for six contrasting soil types in Waikato and Northland, New Zealand, with implications for monitoring. *New Zealand Journal of Agricultural Research* 43: 559-567.

WRC (2012) *Waikato Regional Council dairy winter grazing survey 2011*. Venus Research and A Davies. Waikato Regional Council Technical Report 2012/30.

In: Greenhouse Gas Emissions … ISBN: 978-1-53611-100-2
Editors: Jiafa Luo and Yan Li

Chapter 2

GASEOUS EMISSIONS FROM DAIRY MANURE COLLECTION IN ANIMAL CONFINEMENT FACILITIES

Jiafa Luo[1,2,*], Yan Li[1], Stuart Lindsey[1,2], Benjamin Woodward[3], Xinhao Gao[1], Yongping Jing[1] and Luji Bo[1]

[1]Institute of Agricultural Resources and Environment, Shandong Academy of Agricultural Sciences, Jinan, China
[2]AgResearch, Hamilton, New Zealand
[3]AgResearch, Mosgiel, New Zealand

2.1. INTRODUCTION

Intensification of dairy production worldwide has led to environmental concerns, largely regarding the disposal and management of dairy cow effluent/manure. Animal effluent has a high nutrient content and can cause waterway eutrophication if excess nutrients reach waterways (Rotz 2004). Effluent is also considered an important source of greenhouse gas (GHG) emissions, including nitrous oxide (N_2O) and methane (CH_4). Effluent is also a source of ammonia (NH_3) which is a secondary source of GHGs, and acidifies

* Corresponding author: Dr Jiafa Luo, AgResearch, 10 Bisley Road, Private Bag 3123, Hamilton 3214, New Zealand. Email: Jiafa.Luo@agresearch.co.nz.

ecosystems (Saggar et al. 2004b). The potential for the leaching of nutrients and production of both GHGs and NH_3 are largest in wet and warm seasons (Ledgard et al. 1996; Luo et al. 2000). This has led to increasing attention being placed upon moving livestock from open-grazing fields to confined areas (Longhurst et al. 2006). Animal confinement systems, if well managed, can decrease soil damage, nutrient leaching and emissions of GHGs and NH_3 from dairy farms. But in order for this to be achieved, an understanding of the different types of off-field grazing systems, the effluent generated within these systems and how the management of this effluent influences GHG and NH_3 emissions is needed.

The excreta produced in the confinement facilities include effluents (farm dairy effluent or FDE), slurries and solids. Classification of these depends upon solid or dry matter content. FDE, the most liquid form of manure (dry matter content of 0-5%), is a mixture of water and animal excreta. Slurries, which have a dry matter of 5-15%, are produced by livestock while in the yards or buildings and may have some bedding materials included. They are able to be pumped or discharged by gravity. Solid manures (dry matter content >15%) are able to be stacked in freestanding heaps without slumping. To reach the required dry matter content, solid manures generally need to be dried or mechanically processed.

Regardless of the type of manure generated, the management of animal excreta within confinement systems has three main stages: excreta collection, storage and application to land. Each of these stages has unique environmental conditions that favour the production of different gases. Thus, each stage needs to be managed independently to minimise GHG and NH_3 emissions and leaching losses. Inefficient management of a single stage may result in large losses, offsetting any reductions made in other stages (Amon et al. 2001). This chapter investigates the scientific literature regarding GHG and NH_3 losses associated with collection of manure. Mitigation practices and technologies for reducing GHG and NH_3 emissions are also included and discussed.

2.2. MECHANISMS OF NH_3, N_2O AND CH_4 PRODUCTION

NH_3, N_2O and CH_4 are gases emitted from excreta. These gasses are produced by different processes, each of which occurs under different environmental conditions.

- Volatilisation of NH_3: Volatilisation is the evaporation of a volatile substance, in this case ammonia. The production of NH_3 starts soon after excretion when urea, the primary form of N in animal urine, is converted to ammonia (Monteny and Erisman 1998). The amount and rate of ammonia volatilisation depend upon the Total Ammoniacal (ammonia plus ammonium), Nitrogen (TAN) content of the manure, temperature, moisture content, the pH of the excreta and other environmental conditions (Hartung and Phillips 1994; Sommer and Hutchings 1995).
- Production of N_2O during nitrification: Microbially mediated transformation of NH_4^+ to NO_3^-. N_2O can be produced during nitrification. Its production depends upon the microorganisms present and environmental conditions. Soil conditions affect the N transformation processes and the ratio of N_2O/NO_3^- (Maag and Vinther 1996).
- Production of N_2O during denitrification: Microbially mediated conversion of NO_3^- to N_2. This anaerobic process produces N_2O. A number of soil factors affect the ratio of N_2O/N_2 produced (Blackmer and Bremner 1978; Firestone and Davidson 1989; Firestone et al. 1980; van Cleemput 1998; Weier et al. 1993).
- Production of CH_4: CH_4 is produced during the breakdown of organic matter under anoxic conditions. Organic compounds are degraded by acid-producing bacteria to simpler organic compounds such as volatile fatty acids. After this initial degradation the resulting compounds can be utilised by CH_4 producing bacteria (methanogens) under anoxic conditions (Hellmann et al. 1997).

2.3. Diet and Feeding

Low-protein feed, including maize silage, can reduce the amount of urine N excreted and increase the efficiency of nutrient transformation into product (Van Vuuren et al. 1993). Condensed tannins (CT) in feed derives more of the N that would normally be excreted in urine towards dung (Misselbrook et al. 2005), and N_2O emissions from dung are lower than from urine (van der Weerden et al. 2011). In animal experiments, CT in the diet of sheep (Carulla et al. 2005) and dairy cattle (Grainger et al. 2009) reduced urinary N excretion. Tannin supplementation also decreased ruminal NH_3 concentration.

Salt supplementation encourages animals to take more water, thereby reducing urine-N concentration (van Groenigen et al. 2005; Ledgard et al. 2007). Hippuric acid in urine can affect N_2O emissions and changing the diet to increase its content could be used as a potential mitigation strategy (Kool et al. 2006).

Other feed additives used to improve feed efficiency often reduced N excretion by 5% or less, but reductions of up to 25% have sometimes been reported (Han et al. 2001).

2.4. Animal Pads

Urea in urine can be rapidly converted into NH_3, which can be emitted into the surrounding air (Monteny and Erisman 1998). Materials used in stand-off pads reduce NH_3 emissions from deposited excreta by: reducing wind velocities over the manure; increasing the diffusion distance; absorbing urea (thereby increasing its resistance to conversion to NH_3); chemically binding ammonia (cation exchange capacity) and providing an additional C source which may aid microbial growth and immobilisation of NH_3 (Misselbrook and Powell 2005). A laboratory study by Luo et al. (2006) comparing zeolite, bark, soil and woodchips found that over a 5 week period 13, 25, 30, and 39% of the 403 g N/m^2 applied as excreta was lost as NH_3 from the zeolite, bark, soil and woodchips, respectively. These losses were attributed to the surface area and CEC (cation exchange capacity) of the materials which enabled retention of greater amounts of NH_3 and NH_4^+ thus reducing NH_3 emissions. However, a study by Dumont et al. (2012) that used a range of different sized woodchips, which therefore had different surface areas and CEC capacities, found no difference between their emissions of NH_3. That study did not find any effect of flooring space allocation or cow feeding management on NH_3 emission. However, they did find a significant correlation between wind velocity and NH_3 emission. Adding to the possible influences on NH_3 emission from different bedding materials, Misselbrook et al. (2005) showed that absorbance capacity of bedding material was more influential on NH_3 emissions than either the CEC or pH of the bedding. They reported that bedding materials of sand, pine shavings, chopped newspaper, chopped corn stalks, recycled solid manures and chopped wheat straw emitted 23, 42, 62, 68, 65 and 55% of the total N applied to each material as NH_3. Direct comparisons between NH_3

emissions from standoff pads and conventional grazing practices have not been made. However, the use of a pad will restrict the area over which the excreta are deposited, which should therefore reduce NH_3 emissions.

Luo et al. (2008) illustrated that a standoff pad with a bedding of carbon material, such as pine bark and sawdust, reduced N_2O emissions by 60% compared to standard winter animal grazing methods. Luo and Saggar (2008) suggested that, similar to soils, environmental conditions on standoff pads would affect their emissions of N_2O. In winter, both standoff pads and soil are likely to be saturated and off-grazing pads are most useful, as soil is susceptible to structural damage and the possibility of GHG emissions and leaching of nutrients from soils is the greatest (Saggar et al. 2004a). It is these saturated conditions that Luo et al. (2008) speculate will reduce the amount of oxygen in the pad flooring which reduces the conversion of NH_4^+ to NO_3^- via nitrification, thereby depleting the NO_3^- available for denitrification. So, although the lack of oxygen creates conditions conducive for denitrification, denitrification rates are limited due to a low availability of NO_3^-. Additionally, if denitrification does occur, the anoxic conditions of the flooring material favour the production of N_2 rather than N_2O (Firestone et al. 1980). However, if the flooring materials were allowed to dry out, increases in N_2O emissions could be expected due to increased rates of nitrification and therefore increased availability of NO_3^- for denitrification process and an increased ratio of N_2O to N_2 emitted from this denitrification compared to that found if the pad had remained saturated.

Luo and Saggar (2008) measured winter CH_4 and N_2O emissions from a dairy cow standoff pad in the Waikato region of New Zealand. N_2O emissions were initially low, but during the 86-day monitoring period emission rates peaked and decreased, presumably attributed to pad material water content and nitrate availability. For the first 2 months, when the cows were on the standoff pad, CH_4 emissions were extremely low. However, after this period emissions increased exponentially. This increase was believed to be related to the accumulation of dung over time, creating anoxic conditions within the pad flooring (Luo and Saggar 2008). Over the 3 month period 3.0 ± 0.5 kg CH_4-C was released: removal of wastes from the floor of the pad during this 3 month period probably would have reduced the total CH_4 emission (Luo and Saggar 2008).

2.5. HOUSING

There is a great variety in animal housing facilities used for sheltering dairy cows. However, the key features of housing that influence the emissions of NH_3, N_2O and CH_4 are the area allocated to each animal, flooring type, the methods of excreta collection and the inclusion of internal manure storage areas. Allocated area can be restricted by the use of tie stalls or free-stalls as this reduces the area over which the excreta is deposited. The collection of excreta after deposition onto the floor is dependent upon the flooring type in the facility. Houses with slatted floors allow the excreta to drop through to a collection bunker below the floor, while solid floored houses rely on either regular removal of excreta using a scraper (e.g., European wintering barns) or inclusion of bedding material (straw, sawdust, bark chips) to absorb the excreta during the winter months until the cows return to the fields in early spring (e.g., deep litter barns).

In a slatted floored, cubical house for dairy cows with a collection pit located under the flooring, the flooring accounted for 60% of NH_3 emission and the remaining 40% came from the storage pit (Ogink and Kroodsma 1996). Emissions from storage pits depend on the difference between the temperature of the pit and that above the slates, as this governs the movement of air in or out of the pit. If the air in the pit is warmer than that found outside, NH_3 is emitted; if it is cooler in the pit, NH_3 is retained (Monteny 2000).

Shapes of solid floors and surface characteristics can affect NH_3 loss (Braam et al. 1997a; Braam et al. 1997b). Washing the floor can further reduce NH_3 loss. A number of floor systems, which enable urine to flow away from dung or have smoother coated floors, could be used to reduce NH_3 losses (Braam and Swierstra 1999; Swierstra et al. 2001).

In deep litter systems, straw or sawdust is added to the pit to reduce the exposed surface area of excreta thereby reducing NH_3 emissions (Table 2.1). The inclusion of litter into pits also provides additional mechanisms to retain TAN through binding to CEC sites within the litter and increasing NH_4^+ immobilisation via microbial uptake (Beauchamp et al. 1982; Groot Koerkamp et al. 1998; Amon et al. 2001; Sommer et al. 2006; Bender and Wood 2007). However, the inclusion of straw or sawdust allows a greater amount of oxygen to diffuse into the surface layers, increasing metabolic activity and increasing pit temperature, resulting in convection currents and increased NH_3 emissions (Braam and Swierstra 1999).

Table 2.1. Ammonia emission factors for cattle buildings (adapted from Sommer et al. 2006)

Building design	Pen design	Emission factor (% of total N)	Emission factor (kg NH_3-N per kg TAN)
Tie stalls	Slurry	3	0.6
Cubicle	Partly slatted floors 0.4 m deep slurry channel	6	0.12
Cubicle	Partly slatted floors 1.2 m deep slurry channel	8	0.17
Solid floor	Deep litter	6	0.12

Zhang et al. (2005) showed that flooring had an effect on the emission of NH_3: solid floors with a smooth surface that were scraped and drained were the most effective at minimising NH_3 emissions from dairy cattle free-stall buildings. Slatted floored buildings emitted more NH_3 than smooth floored buildings, but emissions were lowest when acid was added to the excreta and a scraper was used either in drains or on the floor itself. Misselbrook et al. (2000) noted NH_3 emissions of 34.3 and 17.2 g N lu^{-1} d^{-1} (lu = 500 kg of live weight) from dairy cubicles and litter housing systems, respectively. Neither slatted nor solid flooring, nor the addition of straw to slurries had an effect on N_2O or CH_4 emissions (Amon et al. 2001; Zhang et al. 2005).

Regardless of the flooring type, the greater the area over which excreta is deposited the greater the surface area of excreta exposed to air and therefore the larger the NH_3 emissions. Decreasing the exposed excreta surface area can be achieved by either restricting cow movements or by increasing the density of the housed cows. The restriction of animal movements through tie- or free-stalls reduced N losses compared with either deep litter or feedlot houses (Rotz 2004). Amon et al. (2001) reported emission of NH_3 from tie-stalled dairy cows with a slurry or straw based collection system as 5.7 and 5.8 g NH_3 Lu^{-1} d^{-1}, respectively. In a review of NH_3 emissions from dairy cow buildings, where most of the data was gathered from Europe, Monteny and Erisman (1998) stated that NH_3 emissions from cubical houses were 5-27 g day^{-1} per cow and those from tie-stalls were higher at 20-45 g day^{-1} per cow. However, increasing animal density and restricting animal movements must be done within the limits of animal welfare.

Table 2.2. Nitrogen gas emission factors during manure deposition

Manure deposition site	Gaseous emissions (% of total manure N)		
	NH_3	N_2O	N_2
Stand-off pads[1]			
C-rich material floor	2%	0.01%	0.01%
Inert material floor	2%	0.01%	0.01%
Concrete feed pads[2]	2 gN/cow/hr	65 μg N/cow/h	65 μg N/cow/h
Animal shelters[3]			
C-rich material	7%	0	0
Soil	8%	0	0
No lining material	9%	0	0
		NO_3^- is not available during deposition in animal shelters	NO_3^- is not available during deposition in animal shelters

[1] Data for NH_3 and N_2O were based on the data from our measurements with the stand-off pad at DairyNZ 2004/2005. Some of the data has been published (Luo and Saggar 2008). We have assumed 1:1 ratio between N_2O and N_2.

[2] Data for NH_3 and N_2O were estimated from Ellis et al. (2001), Misselbrook et al. (2001) and Hutching et al. (2001), assuming about 3.5 m^2 for a cow. We have assumed a 1:1 ratio between N_2O and N_2.

[3] Data for NH_3 was estimated from Sommer et al. (2006), Hutching et al. (2001), Webb and Misselbrook (2004). Our measurements from the Northland herd homes indicated that no nitrate was produced from the deposition of manure, therefore we assume there are no N_2O or N_2 gases produced at this stage.

2.6. Manure/Slurry Removal

Removal of manure from floors either by mechanical scraping or flushing with water can have some influence on NH_3 emission (Kroodsma et al. 1993). Flushing can reduce NH_3 emission from animal confinement (Kroodsma et al. 1993; Ogink and Kroodsma 1996) and adding acid or disinfectant can further reduce NH_3 loss (Ogink and Kroodsma 1996; Monteny and Erisman 1998).

Conclusion

Removal of dairy cows from paddocks onto standoff pads and barns may lead to pollution swapping, particularly with respect to GHG emissions, as manure must be managed by collection, storage and application to land. During housing, where manure storage is separate from the housing facility, NH_3 emissions are largely affected by surface area over which the excreta is deposited, the collection methods and, where bedding materials are used, the type of bedding material. N_2O and CH_4 emission from housing that is strongly influenced by the frequency of collection, with long the durations between collection events increasing the likelihood of anoxia and CH_4 emissions and increasing the risk of coupled nitrification/denitrification resulting in N_2O emissions.

From the above review of the literature and our limited measurements from stand-off pads and animal shelters, we have suggested values for nitrogen losses from animal confinements and off-paddock facilities (Table 2.2) for incorporation into nutrient budget models and the system analyses (Chapters 8 and 10). These values are based on knowledge of the literature, however, in the future these may vary, as farm systems change and knowledge improves.

ACKNOWLEDGMENTS

The authors acknowledge funding from the Taishan Scholars Program for Overseas Experts and the Ministry for Primary Industries of New Zealand (MPI) through the Sustainable Land Management and Climate Change (SLMACC) programme.

REFERENCES

Amon B., Amon T., Boxberger J. and Alt C. (2001) Emissions of NH_3, N_2O and CH_4 from dairy cows housed in a farmyard manure tying stall (housing, manure storage, manure spreading). *Nutrient Cycling in Agroecosystems* 60: 103-113.

Beauchamp E., Kidd G. and Thurtell G. (1982) Ammonia volatilization from liquid dairy cattle manure in the field. *Canadian Journal of Soil Science* 62: 11-19.

Bender M.R. and Wood C.W. (2007) Above and Below Ground Measurements of Greenhouse Gases from Swine Effluent Amended Soil. *Communications in Soil Science and Plant Analysis* 38: 2479-2503.

Blackmer A.M. and Bremner J.M. (1978) Inhibitory effect of nitrate on reduction of N_2O to N2 by soil microorganisms. *Soil Biology and Biochemistry* 10: 187-191.

Braam C. R. and Swierstra D. (1999) Volatilization of Ammonia from Dairy Housing Floors with Different Surface Characteristics. *Journal of Agricultural Engineering Research* 72: 59-69.

Braam C.R., Ketelaars J.J.M.H. and Smits M.C.J. (1997a) Effects of floor design and floor cleaning on ammonia emission from cubicle houses for dairy cows. *Netherlands journal of Agricultural Science* 45: 49-64.

Braam C.R., Smits M.C.J., Gunnink H. and Swierstra D. (1997b) Ammonia emission from a double-sloped solid floor in a cubicle house for dairy cows. *Journal of Agricultural Engineering Research* 68: 375-386.

Carulla J. E., Kreuzer M., Machmüller A. and, Hess H. D. (2005) Supplementation of Acacia mearnsii tannins decreases methanogenesis and urinary nitrogen in forage-fed sheep. *Australian Journal of Agricultural Research* 56: 961-970.

Dumont P.A., Chadwick D.R., Misselbrook T.H., Robinson J.S., Smith K.A., Sagoo E., Camp V., Murray R., French P., Hill R.A. and Scott A. (2012) Effluent quality and ammonia emissions from out-wintering pads in England, Wales and Ireland. *Agriculture, Ecosystems and Environment* 160: 82-90.

Ellis S., Webb J., Misselbrook T. and Chadwick, D. (2001) Emissions of ammonia (NH_3), nitrous oxide (N_2O) and methane (CH_4) from a dairy herd in the UK. *Nutrient Cycling in Agroecosystems* 60: 115-122.

Firestone M.K. and Davidson E.A. (1989) Microbiological basis of NO and N_2O production and consumption in soil. In: *Exchange of trace gases between terrestrial ecosystems and the atmosphere*. Eds. M.O. Andreae and D.S. Schimel. John Wiley and Sons, Berlin.

Firestone M.K., Firestone R.B. and Tiedje J.M. (1980) Nitrous oxide from soil denitrification - factors controlling its biological production. *Science* 208: 749-751.

Grainger C., Clarke T., Auldist M. J., Beauchemin K. A., McGinn S. M., Waghorn G. C. and Eckard R. J. (2009) Potential use of Acacia mearnsii condensed tannins to reduce methane emissions and nitrogen excretion from grazing dairy cows. *Canadian Journal of Animal Science* 89: 241-251.

Groot Koerkamp P.W.G., Metz J.H.M., Uenk G.H., Phillips V.R., Holden M.R., Sneath R.W., Short J.L., White R.P.P., Hartung J., Seedorf J., Schröder M., Linkert K.H., Pedersen S., Takai H, Johnsen J.O. and Wathes C.M. (1998) Concentrations and Emissions of Ammonia in Livestock Buildings in Northern Europe. *Journal of Agricultural Engineering Research* 70: 79-95.

Han I.K., Piao X.S. and Li D. (2001) Feeding and management system to reduce environmental pollution in swine production. *Asian-Australasian Journal of Animal Sciences* 14: 432-444.

Hartung J. and Phillips V.R. (1994) Control of gaseous emissions from livestock buildings and manure stores. *Journal of Agricultural Engineering Research* 57: 173-189.

Hellmann B., Zelles L., Palojarvi A. and Bai Q. (1997) Emission of climate-relevant trace gases and succession of microbial communities during open-windrow composting. *Applied and Environmental Microbiology* 63: 1011-1018.

Hutchings N.J., Sommer S.G., Andersen J.M. and Asman W.A.H. (2001) A detailed ammonia emission inventory for Denmark. *Atmospheric Environment* 35: 1959-1968.

Kool D. M., Hoffland E., Hummelink E. W.J. and van Groenigen J. W. (2006) Increased hippuric acid content of urine can reduce soil N_2O fluxes. *Soil Biology and Biochemistry* 38: 1021-1027.

Kroodsma W., Huis in 't Veld J.W.H. and Scholtens R. (1993) Ammonia emission and its reduction from cubicle houses by flushing. *Livestock Production Science* 35: 293-302.

Ledgard S.F., Selvarajah N., Jenkinson D. and Sprosen M.S. (1996) Groundwater nitrate levels under grazed dairy pastures receiving different rates of nitrogen fertiliser. In: *Recent developments in understanding chemical movements in soil.* Ed. L P Currie LD. pp 229-236. Fertiliser and Lime Research Centre, Massey University.

Ledgard S.F., Welten B.G., Menneer J.C., Betteridge K., Crush J.R. and Barton M.D. (2007) New nitrogen mitigation technologies for evaluation in the Lake Taupo catchment. *Proceedings of the New Zealand Grassland Association* 69: 117-121.

Longhurst R.D., Luo J., O'Connor M.B. and Pow T. (2006) Herd Homes: a new innovation in dairying to help achieve the 4% per annum production goals. *Proceedings of the New Zealand Grassland Association* 68: 309-313.

Luo J. and Saggar S. (2008) Nitrous oxide and methane emissions from a dairy farm stand-off pad. *Australian Journal of Experimental Agriculture* 48: 179-182.

Luo J., Donnison A., Ross C., Ledgard S. and Longhurst B. (2006) Herd Homes: nutrient management and farmer perceptions of performance. *Proceedings of the New Zealand Grassland Association* 68: 315-320.

Luo J., Ledgard S. and Lindsey S. (2008) A test of a winter farm management option for mitigating nitrous oxide emissions from a dairy farm. *Soil Use and Management* 24: 121-130.

Luo J., Tillman R.W. and Ball P.R. (2000) Nitrogen loss through denitrification in a soil under pasture in New Zealand. *Soil Biology and Biochemistry* 32: 497-509.

Maag M. and Vinther F.P. (1996) Nitrous oxide emission by nitrification and denitrification in different soil types and at different soil moisture contents and temperatures. *Applied Soil Ecology* 4: 5-14.

Misselbrook T.H. and Powell J.M. (2005) Influence of bedding material on ammonia emissions from cattle excreta. *Journal of Dairy Science* 88: 4304-4312.

Misselbrook T.H., Powell J.M., Broderick G.A. and, Grabber J.H. (2005) Dietary manipulation in dairy Cattle: Laboratory experiments to assess the influence on ammonia emissions. *Journal of Dairy Science* 88: 1765-1777.

Misselbrook T.H., van der Weerden T.J., Pain B.F., Jarvis S.C., Chambers B.J., Smith K.A., Phillips V.R. and Demmers T.G.M. (2000) Ammonia emission factors for UK agriculture. *Atmospheric Environment* 34: 871-880.

Misselbrook T.M., Webb J., Chadwick D.R., Ellis S. and Pain, B.F. (2001) Gaseous emissions from outdoor concrete yards used by livestock. *Atmospheric Environment* 35: 5331-5338.

Monteny G. (2000) Modelling of ammonia emissions from dairy cow houses. Wageningen University, Wageningen, The Netherlands.

Monteny G.J. and Erisman J.W. (1998) Ammonia emission from dairy cow buildings: a review of measurement techniques, influencing factors and possibilities for reduction. *Netherlands Journal of Agricultural Science* 46: 225-247.

Ogink N. and Kroodsma W. (1996) Reduction of ammonia emission from a cow cubicle house by flushing with water or a formalin solution. *Journal of Agricultural Engineering Research* 63: 197-204.

Rotz C. (2004) Management to reduce nitrogen losses in animal production. *Journal of Animal Science* 82: E119-E137.

Saggar S., Andrew R.M., Tate K.R., Hedley C.B., Rodda N.J. and Townsend J.A. (2004a) Modelling nitrous oxide emissions from dairy-grazed pastures. *Nutrient Cycling in Agroecosystems* 68: 243-255.

Saggar S., Bolan N.S., Bhandral R., Hedley C.B. and Luo J. (2004b) A review of emissions of methane, ammonia, and nitrous oxide from animal excreta deposition and farm effluent application in grazed pastures. *New Zealand Journal of Agricultural Research* 47: 513-544.

Sommer S.G. and Hutchings N. (1995) Techniques and strategies for the reduction of ammonia emission from agriculture. *Water, Air and Soil Pollution* 85: 237-248.

Sommer S.G., Zhang G.Q., Bannink A., Chadwick D., Misselbrook T., Harrison R., Hutchings N.J., Menzi H., Monteny G.J., Ni J.Q., Oenema O. and Webb J. (2006) Algorithms determining ammonia emission from buildings housing cattle and pigs and from manure stores. *Advances in Agronomy* 89: 261-335.

Swierstra D., Braam C. R. and Smits M. C. (2001) Grooved floor system for cattle housing: Ammonia emission reduction and good slip resistance. *Applied Engineering in Agriculture* 17: 85-90.

van Cleemput O. (1998) Subsoils: chemo- and biological denitrification, N_2O and N_2 emissions. *Nutrient Cycling in Agroecosystems* 52: 187-194.

van der Weerden T.J., Luo J., de Klein C.A.M., Hoogendoorn C.J., Littlejohn R.P. and Rys G.J. (2011) Disaggregating nitrous oxide emission factors for ruminant urine and dung deposited onto pastoral soils. *Agriculture, Ecosystems and Environment* 141: 426-436.

van Groenigen J.W., Velthof G.L., van der Bolt F.J.E., Vos A. and Kuikman P.J. (2005) Seasonal variation in N_2O emissions from urine patches: Effects of urine concentration, soil compaction and dung. *Plant and Soil* 273: 15-27.

Van Vuuren A.M., Van Der Koelen C.J., Valk H. and De H. (1993) Visser Effects of Partial Replacement of Ryegrass by Low Protein Feeds on Rumen Fermentation and Nitrogen Loss by Dairy Cows. *Journal of Dairy Science* 76: 2982-2993.

Webb J. and Misselbrook T.H. (2004) A mass-flow model of ammonia emissions from UK livestock production. *Atmospheric Environment* 38: 2163-2176.

Weier K.L., Doran J.W., Power J.F. and Walters D.T. (1993) Denitrification and the dinitrogen nitrous-oxide ratio as affected by soil-water, available carbon, and nitrate. *Soil Science Society of American Journal* 57: 66-72.

Zhang G., Strøm J.S., Li B., Rom H.B., Morsing S., Dahl P. and Wang C. (2005) Emission of ammonia and other contaminant gases from naturally ventilated dairy cattle buildings. *Biosystems Engineering* 92: 355-364.

In: Greenhouse Gas Emissions … ISBN: 978-1-53611-100-2
Editors: Jiafa Luo and Yan Li

Chapter 3

Gaseous Emissions from Dairy Manure Storage

Jiafa Luo[1,*], Tony van der Weerden[1], Yan Li[2], Cecile de Klein[1], Emily Thom[1], Yongping Jing[2], Yingpeng Zhang[2] and Luji Bo[2]

[1]AgResearch, New Zealand

[2]Institute of Agricultural Resources and Environment, Shandong Academy of Agricultural Sciences, Jinan, China

3.1. Introduction

Several management systems require a separate storage facility for effluent and manure, following collection and prior to land application. Slurries, with solid content ranging from 5 to 15%, are stored in weeping walls, lagoons or ponds. Manures, containing a solid content above 15%, are generally stored as a stack. Storage ponds and lagoons generally consist of aerobic and anaerobic storage areas. The effluent or slurry is initially stored anaerobically then moved into an aerobic pond. The emission of ammonia (NH_3), nitrous oxide (N_2O) and methane (CH_4) from ponds and lagoons is affected by pre-treatment, surface crust, loading methods and environmental factors such as temperature and wind speed (Misselbrook et al. 2001).

[*]Corresponding author: Dr Jiafa Luo, AgResearch, 10 Bisley Road, Private Bag 3123, Hamilton 3214, New Zealand. Email: Jiafa.Luo@agresearch.co.nz.

Storage of solid manure occurs in either heaps or stacks. Aerobic composting or anoxic decomposition can occur during storage. Losses of NH_3, N_2O and CH_4 from solid manure stores are dependent upon the predominance of aerobic composting and anoxic decomposition, storage method, temperature, manure origin, precipitation and whether or not the store is covered.

3.2. Ammonia

The emission of NH_3 is greatest during the initial storage period of fresh slurries or when fresh manure is added to the top of a storage area. Aguerre et al. (2012b) found rates of 852 ± 404 mg NH_3-N h^{-1} when fresh slurries were applied to a storage lagoon but these emissions had decreased by 76% after 7 days of storage and were negligible after 35 days. Similarly, the greatest rate of NH_3 emission from slurry and manure collected from non-lactating dairy cows was during the first few months of storage.

The effects of bedding material type on NH_3 emission from stored slurries have been investigated by several workers. Misselbrook et al. (2005) investigated different cow house bedding materials for reducing NH_3 emissions during slurry storage and found that bedding materials had no effect. Average NH_3 emissions across all treatments during the 80 days of monitoring were 2.3 g N m^{-2} d^{-1}. However, the addition of straw was found by Aguerre et al. (2012a) to have an effect on NH_3 emission - it reduced total emission from 0.47 g m^{-2} to 0.34 g m^{-2}. van der Weerden et al. (2013) found that the addition of straw and sawdust significantly reduced the rate of NH_3 emission when the slurries had been stored for 7 months, with the NH_3 emission factor decreasing with increasing straw and sawdust additions. High sawdust additions resulted in the largest reductions, with the NH_3 emissions factors reduced to 2.8% of the slurry total N content (van der Weerden et al. 2013).

Increased slurry dry matter content can aid the formation of surface crusts which develop over a few days or weeks of storage and decrease the emission of NH_3 from stored slurries (Sommer et al. 1993; Misselbrook et al. 2005; Wood et al. 2012). Sommer et al. (1993) demonstrated that emissions of NH_3 from un-crusted dairy cow slurries were 3-5 g NH_3-N m^{-2} and these emissions were reduced by 80% when a surface crust was present. Similar findings were presented by Misselbrook et al. (2005), however, the reductions in NH_3 emissions were only 50%. The discrepancy in the per cent reduction in NH_3 emissions may be due to Misselbrook et al. (2005) allowing a “skin” to form upon the surface of the control slurry, whereas Sommer et al. (1993) prevented this by stirring their control. The crusted treatments from Misselbrook et al. (2005) had NH_3 emissions ranging from 1.7 – 8.6 g N m^{-2} d^{-1}. Surface crusts are formed either by the slurries themselves as bubbles transport particles to the surface (Misselbrook et al. 2005) or can be aided by the addition of straw or other materials to the surface of the slurry. Regardless of how they are formed, crusts provide a physical barrier between the air and the slurry surface, restricting NH_3 emissions to the rate that NH_3 diffuses through the crust (VanderZaag et al. 2008). A reduction in the pH of the manure after crust formation further reduces NH_3 emissions (Xue et al. 1999; Aguerre et al. 2012a). Additionally, the oxic conditions present within a surface crust allow for populations of NH_3 oxidising bacteria inside the crust which further decreases NH_3 emissions (Hansen et al. 2009). But these oxic crust conditions that decrease NH_3 emissions increase N_2O emissions by creating an oxic/anoxic interface within the crust and allowing coupled nitrification/ denitrification to occur (Aguerre et al. 2012a).

The rate of NH_3 emission from stored solid manures is partly dependent upon the length of the storage period (Sommer 2001; Chadwick 2005; Balsari et al. 2007) and the availability of oxygen to the stored manure (Kirchmann and Lundvall 1998). During the initial period of storage, high rates of metabolism within the manure increase manure temperatures which promotes emissions of NH_3 (Chadwick 2005). Temperatures inside stacked manure heaps that decompose aerobically are greater than those in anaerobic stacks (Amon et al. 2001) driving NH_3 emissions to be greater in aerobic stacks. Turning an aerobically decomposing stack increased NH_3 emissions from 643.3 to 670.5 g of NH_3 t^{-1} FM (FM = farm yard manure) but this effect was only found in summer (Amon et al. 2001). Large reductions in NH_3 emissions can be achieved by storing manures anaerobically. Anaerobic storage reduces rates of metabolism within the stack which reduces manure temperature. The compaction and covering of solid manure was used by Chadwick (2005) as a practical means of reducing the diffusion of O_2 into the manure stores and increasing anaerobic conditions within the stack. In this study conventionally stacked un-compacted and un-covered manure heaps lost 4.5% of total N content, compared to 0.3% of initial N content from compacted and covered stacks (Chadwick 2005). Additionally, Dewes (1996) found that increasing density of farm yard manure reduced NH_3 emission.

Additives have been tested to reduce the NH_3 emission from stored manure with various results (McCrory and Hobbs 2001). The additives include digestive additives, acidifying additives, and adsorbents. Acidifying additives reduce manure pH, thereby reducing NH_3 emission. Potential treatments include acids, base precipitating salts, and labile C (McCrory and Hobbs 2001).

Physical-covering techniques can be used to reduce N loss and odour from slurry storage (Sommer et al. 1993; Hartung and Phillips 1994). The cover includes permanent lid, floating foil or plastic, peat, oil, straw, or polystyrene spheres.

3.3. Nitrous Oxide

Nitrous oxide emission factors from stored manures range from 0.1 to 9.8%, depending on storage methods, periods and seasons (Table 3.1). N_2O emissions from slurries are low during the initial storage period, particularly if no crust is present. After crust formation N_2O emissions increase unless the crusts have a high moisture content (Sommer et al. 2000; Aguerre et al. 2012a). The low N_2O emissions from non-crusted slurries are due to the anoxia within the pond or lagoon inhibiting the bacteria that produce NO_3^- (VanderZaag et al. 2008), however the addition of straw or sawdust increases aeration by allowing greater diffusion of O_2 into the slurry, which increases nitrification rates. This in turn increases the supply of NO_3^- to anaerobic areas within the slurry, fuelling denitrification and increasing N_2O emissions.

Table 3.1. The percentage loss of total N content as N_2O (emission factors) from dairy manures stored as solids under different treatments

Study	Aeration or static composting	Treatment	% of total N emitted as N_2O
Sommer (2001)	Static	Compacted	0.3
		Cut	0.1
		Covered	0.2
		Untreated	0.1
Szanto et al. (2007)	Turned		0.5
	Static		9.8
Amon et al. (2001)	Active aeration	Summer	0.36
	Static	Summer	0.57
	Active aeration static	Winter	0.45 0.88

The presence of surface crusts increase N_2O emission but this increase is affected by the characteristics of the crust. Hansen et al. (2009) noted that the moisture content of the surface crust affected cumulative N_2O emission. For example, Berg et al. (2006) was not able to detect N_2O emissions from pig slurries with a moist surface crust. More specifically, Hansen et al. (2009) found that N_2O emissions increased from 0.26 ± 0.05 g m^{-2} when no crust was present to 28.06 ± 2.68 g m^{-2} with a dry crust; 7.79 ± 1.7 g m^{-2} with a moderately wet crust and 4.0 ± 1.52 g m^{-2} with a wet crust. Surface crusts influence N_2O dynamics by creating an oxygenated surface layer where nitrification can occur producing N_2O and NO_3^- (Hansen et al. 2009). Thus, the size of the oxic surface layer in the surface crust, which is regulated by the moisture content of the crust, influences the rate of N_2O and NO_3^- production (Hansen et al. 2006). Furthermore, high water contents in the surface crust dilute NH_4^+ concentrations, therefore reducing rates of nitrification by limiting substrate availability (Sommer et al. 2000).

The addition of materials to aid in crust formation or to increase the DM content or C:N ratio of slurries influences N_2O emissions. Crusts formed from straw or leca pebbles had lower N_2O emissions, with a delayed onset, when compared to crusts that formed naturally. The reason for this delay was unclear (Sommer et al. 2000). Berg et al. (2006) showed that surface crusts made from different materials emit different amounts of N_2O; emissions ranged from 0.348 ± 0.554 to 2.60 ± 2.85 mg N_2O m^3. In a New Zealand study, the addition of straw and sawdust to slurries to increase their dry matter content was shown to affect N_2O emission from stored slurries (van der Weerden et al. 2013). Either low or high rates of sawdust addition increased rates of N_2O emissions compared to the low or high rates of straw addition or excreta alone. Furthermore, after 7 months of storage these manures were mixed to simulate the emptying of storage facilities. The aeration caused by mixing resulted in the largest N_2O flux, measured from the low sawdust addition treatment, followed by the high sawdust addition treatment. Mixing had little effect on N_2O emissions from slurries where straw was added (van der Weerden et al. 2013).

Similar to NH_3 emissions, the emission of N_2O from stacked solid manures is also related to the diffusion of oxygen into stacks. Sommer (2001) found that over a period of 132 days emissions of N_2O accounted for between 0.1and 0.3% of total N in the manure. Compaction and covering of manures can reduce N_2O emissions by decreasing the penetration of oxygen into the stack and therefore reducing the opportunity for coupled nitrification/ denitrification (Sommer 2001). Compaction was not found to have such clear cut results by Chadwick (2005). They reported N_2O emission factors of 2.3 ± 0.28% from conventional heaps and 2.1 ± 0.66% from compacted heaps. In aerated (non-compacted) stacks, increasing stack density may increase N_2O emissions by increasing the amounts of anoxic microsites within the stack (Webb et al. 2012). Aerated stacks may be turned monthly to reduce the occurrence of anoxic microsites and emissions of N_2O (Szanto et al. 2007).

The close proximity of aerobic and anaerobic zones in solid stacked manures provides conditions where coupled nitrification/denitrification processes can occur, resulting in N_2O emissions. Typically, between <1 and 4.3% of N in either cattle or pig solid manure is lost as N_2O (Chadwick et al. 2011), but losses as large as 9.8% have been reported (Webb et al. 2012). The N_2O emissions from solid manures occurs after a few weeks of storage when stack temperatures have decreased below approximately 50°C as nitrifying and denitrifying bacteria are thermophilic (Hellmann et al. 1997).

3.4. Methane

Methanogenesis requires anaerobic conditions and a carbon source, while the rate of CH_4 production from slurries and solid manures is affected by temperature and other environmental factors (Chadwick et al. 2011). Losses of between 0.4 and 9.7% of total carbon have been observed for solid manures (Chadwick et al. 2011). Both total solids content and slurry temperature at 30 cm below the surface have been positively correlated with CH_4 emission from stored slurries (Sommer et al. 2000; Wood et al. 2012). Similarly, the higher oxygen content found in the crusts of some stored slurries is believed to reduce CH_4 emissions regardless of temperature (Husted 1994; Sommer et al. 2000).

Sommer et al. (2000) documented emissions of CH_4 from stored slurries ranging from <0.01 to 1.4 or 0.7 g CH_4-C m^{-3} h^{-1}. Husted (1994) monitored CH_4 emission for a cattle slurry over a year and found 38% higher CH_4 emissions when slurries were uncovered compared to slurries that had a naturally formed surface crust. A decrease in CH_4 emission with naturally formed crusts has also been reported by others (Laguee et al. 2005; Petersen et al. 2005). The effects of artificial crusts, constructed with straw or crop residues is not as clear; some studies have found reductions in CH_4 emission of up to 90% (Laguee et al. 2005; Petersen et al. 2005) while others have noted 250% increases in CH_4 emissions (Cicek et al. 2004). This variation may be due to how surface crusts are reducing CH_4 emissions. It is thought that crusts on slurries act as chemical barriers to CH_4 emissions. Evidence for this was presented by Petersen et al. (2005) who showed the oxidation of CH_4 to CO_2 was occurring in naturally occurring and artificial surface crusts of cattle slurries. Hansen et al. (2009) was unable to find a population of methanotrophs large enough to account for the observed reduction in CH_4 emission in crusted cattle slurries when compared to un-crusted slurries. Therefore, crusts may be acting as both chemical and physical barriers to CH_4 emissions. Thus, if cracks appear due to drying, mixing or preparation for land application, gases stored under the crust may be released.

In addition to aiding crust formation, materials may also be added to slurries and manure to increase their dry matter content and increase the C:N ratio and stack aeration (Yamulki 2006). The effect of such additions on CH_4 emissions depends on the type and amount of substrate being added (van der Weerden et al. 2013). Over a 7-month period, the addition of straw at high or low rates to a slurry had little effect on the CH_4 emission factor of the stored manure (0.77% and 0.86% of excreta-C content in the high and low rate additions, respectively, compared to 0.61% for the non-amended slurry). In contrast, when a high rate of sawdust was added to slurries, CH_4 emissions were suppressed entirely, possibly as a result of the inhibition of methanogenic fermentation of the organic material due to the low pH (4.3) of the sawdust (van der Weerden et al. 2013).

Reductions can also be achieved by maximising either aerobic or anaerobic conditions within the stack (Chadwick et al. 2011). Hansen et al. (2006) showed that the rates of CH_4 emission from un-covered manure storages were 1.6 kg t^{-1} or 1.3% of the initial C content, while covering the stacks reduced the rate of CH_4 emission to 0.2 kg t^{-1} or 0.17% of the initial C content. Compaction of stacks maximises anaerobic conditions within the stack, thereby reducing CH_4 emissions by reducing both in-stack metabolic rates and temperature. However, Sommer (2001) found that emissions of CH_4 were 48.06, 44.05, 9.36 and 7.56 g CH_4-C t^{-1} from compacted, cut and mixed, covered and untreated manure solids, respectively. Therefore, the effects of compaction seem to be variable, but this result may be related to temperature within the manure stack, which was highest in the compacted and cut and mixed treatments. Alternatively, frequent turning of heaps will reduce anoxic pockets within the heaps. This was shown to reduce CH_4 losses to approximately 0.5% of total C content (Amon et al. 2001; Amon et al. 2006).

A biofilter capable of removing most of the CH_4 produced from dairy farm waste ponds has been developed (Tate et al. 2012). Active CH_4-eating bacteria (methanotrophs) in the biofilter convert the CH_4 into biomass and some CO_2; NH_3 is also consumed. The concept suggests that a volcanic soil exposed to high CH_4 fluxes could remove these waste emissions (Tate et al. 2012). This concept was later confirmed using a biofilter where almost all the CH_4 emissions produced from an average (ca. 450 cows) dairy farm waste pond could be removed (Pratt et al. 2012b). This biofilter has operated successfully with minimal maintenance on an effluent pond, over more than three years, removing up to 16 g CH_4 m^{-3} h^{-1} (Pratt et al. 2012a). An advanced biofilter design which floats on livestock slurry is being tested on a commercial dairy farm, with similar high efficiencies for CH_4 removal being achieved (Pratt et al. 2013).

CONCLUSION

When confined animals void their urine and dung on stand-off/feed pads or in houses, sheds and barns, it is generally stored for some time before subsequent application to the land. There are wide variations in animal manure/slurry storage and management systems that include differences in manure collection methods, amendments with litter or other materials, length of the storage period, the type of manure cover used during storage. A major factor affecting nitrogen loss is the manure dry matter (DM) content. Other manure characteristics that affect nitrogen loss include total nitrogen concentration, ammoniacal nitrogen concentration, and pH. Environmental factors, such as ambient temperature, wind velocity, and solar radiation, also affect the rate of loss from open storage. As reviewed above, the literature gives a wide range of emission factors from different storage systems. Table 3.2 gives "average" nitrogen emission factors during manure storage, which may be included in nutrient budget models and system analyses (Chapters 8 and 10). NH_3 gaseous emission factors given are mostly from those published by Oenema and Tamminga (2005). N_2O emission factors are mostly from those published by Chadwick (2005) and IPCC (2006) default values. N_2 emission factors are mostly from those published by Oenema and Tamminga (2005). Some proportions of nutrients are leached from open solids storage, but these losses depend on the amount of rainfall. Partitioning of nutrients between solid and liquid phases of stored effluent is dependent on storage time in ponds and whether stirring happens before land application (Luo and Longhurst 2008). More studies are needed to quantify gaseous emissions by conducting field measurements, so that more accurate data can be incorporated into the total GHG footprint analysis and a fair comparison between all-year grazing and off-grazing systems can be made.

As exposure of manure slurries to the atmosphere during storage influences the emission of GHGs and NH_3 in this intermediate stage of manure management, reducing the exchange of gases between slurries and the atmosphere can be accomplished by crusting (artificial or natural) or by using sealed enclosures, while solid manures can be covered and compacted. Reducing gaseous exchanges will reduce NH_3 volatilisation and promote complete anaerobic conditions which will reduce nitrification and nitrate-N production, while promoting complete denitrification (to benign N_2). However, these anaerobic conditions may promote net CH_4 production, with emissions being dependent upon storage length and temperature.

Table 3.2. Nitrogen gas emission factors during manure storage

% of total N				
	Gaseous emissions			Liquids leached from solids
	NH_3	N_2O	N_2	
Holding ponds	15%	0.1%	5%	
Anaerobic/aerobic ponds	30%		20%	
Advanced ponds	30%	1%	32%	
Slurries	60%/year *	0.1%/year *	2%/year*	
Covered solids	30%/year *	1%/year *	10%/year *	0%
Open solids	30%/year *	1%/year *	10%/year *	5% N/year, 0% P, 10% K, 5% S, 0% Ca, 0% Mg, 10% Na *

* For slurries and solids, period stored needs to be calculated, i.e., emission rate × months/12.

ACKNOWLEDGMENTS

The authors acknowledge funding from the Ministry for Primary Industries of New Zealand (MPI) through the Sustainable Land Management and Climate Change (SLMACC) programme and the Taishan Scholars Program for Overseas Experts.

REFERENCES

Aguerre M. J, Wattiaux M. A., Hunt T. and Lobos N. E. (2012a) Effect of nitrogen content and additional straw on changes in chemical composition, volatile losses, and ammonia emissions from dairy manure during long-term storage. *Journal of Dairy Science* 95: 3454-3466.

Aguerre M. J., Wattiaux M. A. and Powell J. M. (2012b) Emissions of ammonia, nitrous oxide, methane, and carbon dioxide during storage of dairy cow manure as affected by dietary forage-to-concentrate ratio and crust formation. *Journal of Dairy Science 95*: 7409-7416.

Amon B., Amon T., Boxberger J. and Alt C. (2001) Emissions of NH_3, N_2O and CH_4 from dairy cows housed in a farmyard manure tying stall (housing, manure storage, manure spreading). *Nutrient Cycling in Agroecosystems* 60: 103-113.

Amon B., Kryvoruchko V., Moitzi G., Amon T. and Zechmeister-Boltenstern S. (2006) Methane, nitrous oxide and ammonia emissions during storage and after application of dairy cattle slurry and influence of slurry treatment. *Agriculture, Ecosystems and Environment* 112: 153–162.

Balsari P., Airoldi G., Dinuccio E. and Gioelli F. (2007) Ammonia emissions from farmyard manure heaps and slurry stores-Effect of environmental conditions and measuring methods. *Biosystems Engineering* 97: 456-463.

Berg W., Brunsch R. and Pazsiczki I. (2006) Greenhouse gas emissions from covered slurry compared with uncovered during storage. *Agriculture, Ecosystems and Environment* 112: 129-134.

Chadwick D., Sommer S., Thorman R., Fangueiro D., Cardenas L., Amon B. and Misselbrook T. (2011) Manure management: Implications for greenhouse gas emissions. *Animal Feed Science and Technology* 166-167: 514-531.

Chadwick D. R. (2005) Emissions of ammonia, nitrous oxide and methane from cattle manure heaps: effect of compaction and covering. *Atmospheric Environment* 39: 787-799.

Cicek N, Zhou X, Zhang Q and Tenuta M. (2004) Impact of straw cover on greenhouse gas and odor emissions from manure. *In ASAE Annual International Meeting.* ASAE.

Dewes T. (1996) Effect of pH, temperature, amount of litter and storage density on ammonia emissions from stable manure. *The Journal of Agricultural Science* 127: 501-509.

Hansen M. N., Henriksen K. and Sommer S. G. (2006) Observations of production and emission of greenhouse gases and ammonia during storage of solids separated from pig slurry: Effects of covering. *Atmospheric Environment* 40: 4172-4181.

Hansen R. R., Nielsen D. A., Schramm A., Nielsen L. P., Revsbech N. P. and Hansen M. N. (2009) Greenhouse gas microbiology in wet and dry straw crust covering pig slurry. *Journal of Environmental Quality* 38: 1311-1319.

Hartung J. and Phillips V. R. (1994) Control of gaseous emissions from livestock buildings and manure stores. *Journal of Agricultural Engineering Research* 57: 173-189.

Hellmann B., Zelles L., Palojarvi A. and Bai Q. (1997) Emission of climate-relevant trace gases and succession of microbial communities during open-windrow composting. *Applied and Environmental Microbiology* 63: 1011-1018.

Husted S. (1994) Seasonal variation in methane emission from stored slurry and solid manures. *Journal of Environmental Quality* 23: 585-592.

IPCC (2006) IPCC Guidelines for National Greenhouse Gas Inventories: Volume 4: Agriculture, Forestry and other Land Use. Intergovernmental Panel on Climate Change. Paris, France. Available at: http://www.ipcc-nggip.iges.or.jp/public/2006gl/vol4.htm.

Kirchmann H. and Lundvall A. (1998) Treatment of solid animal manures: identification of low NH3 emission practices. *Nutrient Cycling in Agroecosystems* 51: 65-71.

Laguee C., Gaudet E., Agnew J. and Fonstad T. (2005) Greenhouse gas emissions from liquid swine manure storage facilities in Saskatchewan. *Transactions of the ASAE* 48: 2289-2296.

Luo J. and Longhurst B. (2008) Improved livestock management systems and nutrient losses from manure deposition and treatment – A support document for Overseer nutrient budget model.

McCrory D. F. and Hobbs P. J. (2001) Additives to reduce ammonia and odor emissions from livestock wastes. *Journal of Environmental Quality* 30: 345-355.

Misselbrook T. H., Powell J. M., Broderick G. A. and, Grabber J. H. (2005) Dietary manipulation in dairy Cattle: Laboratory experiments to assess the influence on ammonia emissions. *Journal of Dairy Science* 88: 1765–1777.

Misselbrook T. H., Webb J., Chadwick D. R., Ellis S. and Pain B. F. (2001) Gaseous emissions from outdoor concrete yards used by livestock. *Atmospheric Environment* 35: 5331–5338.

Oenema O. and Tamminga S. (2005) Nitrogen in global animal production and management options for improving nitrogen use efficiency. *Science in China Ser. C Life Science* 48: 871-887.

Petersen S. O., Amon B. and Gattinger A. (2005) Methane Oxidation in Slurry Storage Surface Crusts. *Journal of Environmental Quality* 34: 455-461.

Pratt C., Deslippe J. and Tate K. R. (2012a) Testing a biofilter cover design to mitigate dairy effluent pond methane emissions *Environmental Science and Technology* 47: 526-532.

Pratt C., Walcroft A. S., Deslippe J. and Tate K. R. (2013) CH_4/CO_2 ratios indicate highly efficient methane oxidation by a pumice landfill cover-soil. *Waste Management* 33: 412-419.

Pratt C., Walcroft A. S., Tate K. R., Ross D. J., Roy R., Reid M. H. and Veiga P. W. (2012b) Biofiltration of methane emissions from a dairy farm effluent pond Agriculture, *Ecosystems and Environment* 152: 33–39.

Sommer S. G. (2001) Effect of composting on nutrient loss and nitrogen availability of cattle deep litter. *European Journal of Agronomy* 14: 123-133.

Sommer S. G., Christensen B. T., Nielsen N. and Schjφrring J. (1993) Ammonia volatilization during storage of cattle and pig slurry: effect of surface cover. *The Journal of Agricultural Science* 121: 63-71.

Sommer S. G., Christensen B. T., Nielsen N. and Schjφrring J. (1993) Ammonia volatilization during storage of cattle and pig slurry: effect of surface cover. *The Journal of Agricultural Science* 121: 63-71.

Sommer S. G., Petersen S. O. and Søgaard H. T. (2000) Greenhouse gas emission from stored livestock slurry. *Journal of Environmental Quality* 29: 744-751.

Szanto G. L., Hamelers H. V. M., Rulkens W. H. and Veeken A. H. M. (2007) NH_3, N_2O and CH_4 emissions during passively aerated composting of straw-rich pig manure. *Bioresource Technology* 98: 2659-2670.

Tate K. R., Walcroft A. S. and Pratt C. (2012) Varying atmospheric methane concentrations affect soil methane oxidation rates and methanotroph populations in pasture, an adjacent pine forest, and a landfill. *Soil Biology and Biochemistry* 52: 75–81.

van der Weerden T., Luo J., Dexter M. and Kear M. (2013) Does adding straw or sawdust to cattle excreta reduce ammonia, nitrous oxide and methane emissions during storage? In: Accurate and efficient use of nutrient on farms. (Eds. L. D. Currie and C. L. Christensen). http://flrc.massey.ac.nz/ publications.html. Occasional Report No. 26. Fertilizer and Lime research Centre, Massey University, Palmerston North, New Zealand. 12 pages.

VanderZaag A., Gordon R., Glass V. and Jamieson R. (2008) Floating covers to reduce gas emissions from liquid manure storages: a review. *Applied Engineering in Agriculture* 24: 657-671.

Webb J., Sommer S., Kupper T., Groenestein K., Hutchings N., Eurich-Menden B., Rodhe L., Misselbrook T. and Amon B. (2012) Emissions of ammonia, nitrous oxide and methane during the management of solid manures. In Agroecology and Strategies for Climate Change. Ed. E Lichtfouse. pp 67-107. Springer Netherlands.

Wood J. D., Gordon R. J., Wagner-Riddle C., Dunfield K. E. and Madani A. (2012) Relationships between dairy slurry total solids, gas emissions, and surface crusts. *Journal of Environmental* Quality 41: 694-704.

Xue S., Chen S. and Hermanson R. (1999) Wheat straw cover for reducing ammonia and hydrogen sulfide emissions from dairy manure storage. *Transactions of the ASAE* 42: 1095-1101.

Yamulki S. (2006) Effect of straw addition on nitrous oxide and methane emissions from stored farmyard manures. *Agriculture, Ecosystems and Environment* 112: 140-145.

In: Greenhouse Gas Emissions …
Editors: Jiafa Luo and Yan Li
ISBN: 978-1-53611-100-2

Chapter 4

GASEOUS EMISSIONS AND LAND APPLICATIONS OF SLURRIES AND SOLID MANURES

Yan Li[1,*], Jiafa Luo[1,2], Yongping Jing[1], Yingpeng Zhang[1], Zhaohui Liu[1], Luji Bo[1] and Deshui Tan[1]

[1]Institute of Agricultural Resources and Environment, Shandong Academy of Agricultural Sciences, Jinan, China
[2]AgResearch, Hamilton, New Zealand

4.1. INTRODUCTION

Land application of slurries and solid manures is the final step in the management of effluent generated by off-paddock grazing systems inside animal confinements. The benefits of reduced emissions of ammonia (NH_3), nitrous oxide (N_2O) and methane (CH_4), achieved in the previous stages of manure management in the animal confinements, may be lost in this final stage, lowering the potential economic benefits of applying manures to croplands and creating a source of atmospheric pollution. The re-introduction of slurries and manures into the environment is one of the largest challenges in slurry and manure management due to the high

[*]Corresponding author: Dr Yan Li, Institute of Agricultural Resources and Environment, Shandong Academy of Agricultural Sciences, Jinan 250100, China. E-mail: nkyliyan@126.com.

variation in environmental conditions encountered. These conditions influence the production of NH_3, N_2O and CH_4.

4.2. Ammonia

The characteristics and history of the manure being applied to land will affect emissions of NH_3. In the short term, the amount of Total Ammoniacal Nitrogen (TAN, NH_3 + NH_4^+) contained within the manure determines amount of NH_3 that is available for volatilisation from the applied manure. The amount of TAN contained within solid manures or slurries depends upon: the type of manure, time passed since excretion and how it has been managed during this time, as discussed in Chapters 2 and 3. Solid manures or slurries with a high solid content do not infiltrate quickly into the soil and are therefore susceptible to NH_3 emissions (Pain et al. 1989; Sommer and Olsen 1991). But the relationship between dry matter content and NH_3 emission is not linear as increased dry matter content of a slurry or solid manure will reduce its pH, which in turn, reduces NH_3 emissions during land application (Sommer and Olesen 1991).

The major environmental factors affecting NH_3 volatilisation from land applied slurries or solid manures include wind speed, rainfall and temperature (Thompson et al. 1990; Misselbrook et al. 2005). Formation of a surface crust can reduce emissions from surface applied slurries and solid manures, as the crust serves as a barrier to gaseous emissions. Regular wetting due to water input can affect crust formation (Misselbrook et al. 2005), increasing NH_3 emissions. However, a single rainfall or irrigation event, soon after surface land application of slurries, that deposits approximately 20 mm of water can reduce NH_3 emissions (Misselbrook et al. 2005; Webb et al. 2012). This is due to moving NH_4^+ from the slurry into the soil profile, thereby reducing the TAN content on the soil surface. However, infiltration, and therefore the transporting of NH_4^+ into the soil profile, during rainfall or irrigation events is dependent on soil antecedent moisture conditions. Theoretically, higher temperatures should increase NH_3 emissions; however, Misselbrook et al. (2005) did not find a relationship between NH_3 emissions and temperature. They suggested that other factors, such as soil moisture, humidity, wind speed and rainfall, may have cancelled the influence of temperature on NH_3 emissions. If manures can be incorporated into the soil profile, the TAN is removed from contact with the atmosphere and therefore the potential for NH_3

emissions is reduced. Injection or infiltration of slurries reduces the losses of NH_3 (Rotz 2004; Webb et al. 2010).

Of all the methods used to apply solid manures or slurries to soils, NH_3 losses from broadcast spreading are largest due to the large surface area of manure that is exposed to the atmosphere (Misselbrook et al. 2002; Webb et al. 2010). Misselbrook et al. (2002) applied cattle slurries to pastures using broadcast spreading, trailing shoe and shallow injection. Expressed as a percentage of initial TAN content in the applied slurries, NH_3 losses were 35, 12 and 13%, respectively. NH_3 emissions can be significantly reduced by incorporating surface applied manures into the soil by tillage (Rotz 2004). Mkhabela et al. (2008) compared no-tillage and conventional-tillage systems across two sites and demonstrated that conventional tillage reduced NH_3 emissions by 72-91% in 2003/2004 and in 2004/2005 reductions were 81-83%. These reductions are consistent with data reported by Malgeryd (1998) and Sommer and Hutchings (1995). Most NH_3 is emitted within 4 hours of application to the soils (Webb et al. 2010), therefore incorporation of applied manures must occur immediately following application.

Band spreading and injection of slurries into the soil can reduce NH_3 emissions compared to broadcast spreading. Band spreading resulted in 35% of TAN being emitted as NH_3 compared to 47% with broadcast spreading when manure was applied to a grassland soil (Misselbrook et al. 2002). However, the efficiency of band spreading at reducing NH_3 emission is less if bands have a large surface area i.e., large band widths (Rotz 2004). The lowest NH_3 emissions occur if slurries are injected into the soil - compared to broadcast spreading, a 73% reduction was noted (Misselbrook et al. 2002). The efficiency of injection methods for reducing NH_3 emissions varies due to variations in soil and crop conditions and other operational aspects (Misselbrook et al. 2002).

4.3. Nitrous Oxide

There is some evidence that low losses of NH_3 after field application of solid manures or slurry will lead to high N_2O emissions (Flessa and Beese 2000; Wulf et al. 2002). However, this is not always the case (Sommer et al. 1996; Vallejo et al. 2005). The retention of NH_4^+ in the soil by restricting NH_3 losses provides more substrate for nitrification and subsequent denitrification and the generation of N_2O by these processes. However, NH_4^+ may also be utilised by plants or bound to soil exchange sites and therefore not be available for nitrification or denitrification. Emissions of NH_3 and N_2O are often temporally separated, with N_2O emission delayed after the application of manures (Chadwick et al. 2011). A large range of N_2O emission factors have been reported after the application of manure slurries or solids to land, ranging from <0.1 to 3% (Chadwick et al. 2011). This range in emission factors results from differences in soil conditions and manure composition.

Soil characteristics such as C and N content and texture, as well as more variable soil properties such as water-filled pore space and temperature, interact to influence the emissions of N_2O from land applied manures. Water-filled pore space and temperature vary temporally and between soil types and are both positively related to N_2O emissions from land applied manures (Dobbie et al. 1999). Sommer et al. (1996) found greater emissions from surface applied and injected swine manures when gravimetric soil moisture was 24% (w/w) compared to 10% (w/w). Comparing emission factors from sandy and clay soils, van Groenigen et al. (2004) found that 0.51% of applied N was lost as N_2O from a sandy soil and 1.21% from a clay soil. However, Jarecki et al. (2008) applied swine manure slurry to both a sandy loam and a clay soil and observed larger emission factors from the sandy loam (2.7% of applied N) when compared to the clay soil (1.84% of applied N). The lower emission factors from the clay soil in Jarecki et al. (2008) were suggested to be caused by a greater CEC (cation exchange capacity), which allowed the soil to retain more N in the form of NH_4^+, thus reducing N_2O production in both nitrification and denitrification. The mechanisms behind N_2O emissions in clay and sandy soils are thought to be different: nitrification is suggested to be the source in sandy soils as they remain aerobic after manure application, while denitrification is the dominant mechanism in clay soils due to anaerobic conditions after application (Rochette et al. 2008). The temporal variability in the environmental factors that affect N_2O production in soils, and the different mechanisms thought to be responsible for N_2O production between clay and sandy soils, suggests knowledge of both soil conditions and manure characteristics are necessary for reducing N_2O emissions. One such characteristic of the manure is the C content, which can enhance denitrification rates and the production of N_2O (Chadwick et al. 2000). However, solid manures usually have a lower N content, which lead to lower N_2O emissions than slurries (Chadwick et al. 2011).

The emission of N_2O from applied manure is also affected by the application method. Slurries, which can be broadcast, banded or injected, have largely been shown to have the greatest emission after injection. Ellis et al. (1998) applied cattle slurries by surface application and shallow injection, and observed temporal variation in N_2O emissions, with cumulative losses being greater from injection. After injection, N_2O emission peaked at 34 g N ha^{-1} d^{-1}, decreased to 10 g N ha^{-1} d^{-1} 3 days after injection, but increased again to 13 g N ha^{-1} d^{-1} following a rainfall event. Similar variation was also evident in N_2O emissions from surface applied slurries, with rates peaking at 9 g N ha^{-1} d^{-1} immediately after application, decreasing to 4 g N ha^{-1} d^{-1} the following day but increasing again to 7 g N ha^{-1} d^{-1} following a rainfall event. Thomsen et al. (2010) also found larger emission factors from injection methods compared to banded applications. Emission factors were 2.5 and 1.9% when using injection methods of straight and winged tine, respectively. These values were greater than the 0.5% calculated for surface application using a trailing shoe. However, in the second year of this study the trailing shoe had an emission factor of 0.7%, while the only injection method used, the straight tine, had a lower emission factor of 0.4%. Variation in temperature and soil pH influence the N_2O emission from soils after the application of manures in either slurry or solid form (Chadwick et al. 2011). Rodhe et al. (2006) found emission factors of 1.1% when injecting cattle slurry into a silty clay soil and 0.3% when this slurry was banded onto the same soil using a hose 0.05 m above the soil surface. The method used to apply manures to the soil affects the emission of N_2O after application as do the soil conditions and manure properties.

The effect of application methods on N_2O emissions from land applied solid manure is not clear. In a study over 2 years, Webb and Misselbrook (2004) found emission factors of 0.33% from surface applied cattle solids compared to 0.02 and <0.01 when solids were incorporated into the soils 4 and 24 hours after application, respectively. Similar results were presented for the second year of this study. When surface applying solid swine manure to a sandy loam soil, Thorman et al. (2007) found emission factors were 0.07% from both stored and fresh manures that were surface applied, but when manures were immediately incorporated in the soil, emission factors increased to 0.29% from stored manures and 3.27% from fresh manures. This variation suggests that soil conditions play an influential role in determining the N_2O emissions from land application of solid manures. Thorman et al. (2007) suggested that when soil conditions are conducive to complete denitrification there is a greater likelihood of N_2 being produced via denitrification rather than N_2O.

Urease and nitrification inhibitors (UIs and NIs) can be used to reduce N losses from excreta by inhibiting the N processes of urea hydrolysis and nitrification, respectively (Watson 2000). NIs delay the nitrification process and keep soil N in the more immobile NH_4^+ form in soil. The majority of research indicates that applying UIs to soils with excreta reduce NH_3 volatilisation while applying NIs reduce N_2O emissions. Some studies also show NIs increase NH_3 volatilisation (Davies and Williams 1995; Li et al. 2014). Di and Cameron (2002) found that the NI DCD (dicyandiamide) application following two urine applications reduced N_2O emissions by 82%. The NI 3,4-dimethylpyrazole phosphate (DMPP), applied at a rate of 1 kg ha^{-1}, reduced N_2O emissions by 60% in autumn and by 48% in spring when applied to grassland after slurry application (Merino et al. 2005). Results of using NIs to reduce denitrification are not conclusive (de Klein et al. 1996; Stoeven et al. 2002).

4.4. Methane

Methane emissions occur immediately following manure application to land and then sharply decrease (Chadwick et al. 2000; Rodhe et al. 2006; Sommer et al. 2006). There are two reasons for this: firstly, methanogens are sensitive to O_2 and the diffusion of O_2 into manure after application to land, regardless of the application method, increases when compared to storage. The greater diffusion of O_2 inhibits CH_4 emissions (Chadwick et al. 2011). Secondly, it is thought that most of the CH_4 emitted from land-applied slurries and solid manures is from CH_4 produced before land application and dissolved within the liquid fraction of these manures (Rodhe et al. 2006; Sommer et al. 2006). Chadwick et al. (2000) showed that >90% of the 0.60 g ha^{-1} of CH_4 emitted over a two month period occurred within one day of manure application.

Over time, soils may become CH_4 sinks. Rodhe et al. (2006) reported that over a 7 week period following the application of a cattle slurry to a grassland soil, the soil consumed 362 g CH_4-C ha^{-1}. But whether the soils become sinks for CH_4 may depend upon the application method. Using laboratory incubations to investigate the CH_4 emissions from surface applied and injected cattle slurries, Flessa and Beese (2000) found a flux over a two week period of -12 g ha^{-1} where no slurry had been applied, 2 g ha^{-1} for surface application and 39 g ha^{-1} from injected slurries. This finding is not unexpected as injection in the soil would restrict O_2 diffusion in the manures promoting anaerobic conditions and CH_4 production. However, Sommer et al. (1996) found no significant difference in total CH_4 emission between slurries that were either surface applied, mixed into the top 5 cm of soil or injected into the soil. Correlations have also been noted between the content of volatile fatty acids in the spread manure and CH_4 emitted after manure application (Sommer et al. 1996; Kirchmann and Lundvall 1998).

4.5. Comparisons of Off-Paddock Grazing and Conventional Grazing Systems

Two studies have compared GHG emissions from conventional grazing and off-paddock grazing systems. Both of these studies were completed in New Zealand and compared N_2O emissions, when stand-off pads were used to remove dairy cows during periods of potentially high N_2O emissions from grazed pastures. To compare N_2O emissions from the total farm systems, both of these studies combined measured N_2O fluxes from paddocks and stand-off pads with calculated N_2O emissions from other sources such as effluent storage and land application. Both studies calculated emissions based on the Intergovernmental Panel for Climate Change (IPCC) inventory methodology. Luo et al. (2008) completed a 2 year study in the Waikato Region of New Zealand, using stand-off pads constructed from screened crushed pine bark and /or sawdust (Table 4.1). The stand-off pads were used by cows for 18 hours per day, about 3 months a year, between late autumn and early spring. The use of the stand-off pad significantly reduced N_2O emissions from the grazed pastures from 2.99 to 1.33 kg N_2O-N ha^{-1} in the first year and from 1.22 to 0.28 kg N_2O-N ha^{-1} in the second year. Over the whole farm, the use of stand-off pads reduced emission from 7.7 to 7.0 kg N_2O per hectare of grazed pastures. When compared to the reductions in N_2O emission from the grazed paddocks themselves, this smaller reduction in whole farm emission is due to a 324% increase in N_2O emission, resulting from the land application of effluent collected from the stand-off pad. Similar results were presented by de Klein et al. (2006) who used a stand-off pad on a dairy farm in Southland, New Zealand. Lactating cows were removed from the pasture for 21 hours a day during March, April and May for 3 years. N_2O emissions from the grazed pasture, measured between April and October, were reduced by 57% from 1.4 to 0.6 kg N ha^{-1}. But reductions from the entire farm were only 7%. Again this was due to an increase in N_2O emission resulting from the land application of the collected effluent. An LCA methodology will be used to compare total GHG emissions between different farm systems involving off-paddock facilities (Chapter 8).

Table 4.1. Annual N_2O emissions for whole farm systems (kg N_2O-N ha^{-}) (Data from Luo et al. 2008)

Source	Control	Stand-off
Direct emission (including application of fertilizer and manure and deposition of excreta	5.86	5.25
Indirect emission (including emissions related to leached N and volatilised NH_3 from stand-off pads and fertilizer and excreta)	1.85	1.78
Total	7.7	7.0
Reduction in N_2O emission compared to control (%)		9

CONCLUSION

During land application, manure and slurry exposure to the atmosphere can be reduced by either banding applications or injection into the soil, thus reducing NH_3 emissions. However, this may increase N_2O emissions due to the presence of both aerobic and anaerobic zones in the manure and/soil, but the development of these zones is largely dependent on soil moisture levels at the time of application. Emissions of CH_4 from land applications are believed to be minimal, largely resulting from CH_4 produced during storage.

Each step in the management of animal manures has its own set of physical and environmental conditions affecting the production and emission of GHG, thus each step is generally managed independently. However, it is also important to highlight that management of these manures should be viewed as a continuum, where potential emissions from one stage are directly influenced by the conditions and subsequent emissions that occurred in the previous stages. For example, a livestock waste management system may include an anaerobic process followed by an aerobic process. This would reduce CH_4 emissions, but would be likely to increase N_2O emissions. Therefore, reductions in GHG production at one stage may result in increased production at a later stage, unless all stages are well managed. It is also important to note that there are several GHGs (N_2O, NH_3 and CH_4), and adjusting manure management practices to reduce one gas may indeed increase the emission of another. Therefore, management of these emissions must be done with a holistic view of all the stages involved and with consideration of the total GHG footprint. Due to the complexity and potential interaction between the factors affecting GHG emissions, both within and between management stages, systems based modelling should play an important role in designing manure management systems and assessing potential reductions in emissions when dairy cows are taken off-paddock.

ACKNOWLEDGMENTS

This work was supported by the National Key Technology R&D Program of the Ministry of Science and Technology (2012BAD15B02), the Taishan Scholars Program for Overseas Experts, Special Fund for Agro-scientific Research in the Public Interest (201203050) and Key projects of scientific and technological innovation of Shandong Academy of Agricultural Sciences (2014CXZ01), and Ministry for Primary Industries in the Sustainable Land Management and Climate Change programme.

REFERENCES

Chadwick D., Sommer S., Thorman R., Fangueiro D., Cardenas L., Amon B. and Misselbrook T. (2011) Manure management: Implications for greenhouse gas emissions. *Animal Feed Science and Technology* 166-167: 514-531.

Chadwick D. R., Pain B. F. and Brookman S. K. E. (2000) Nitrous Oxide and Methane Emissions following Application of Animal Manures to Grassland. *Journal of Environmental Quality* 29: 277-287.

Davies D. M. and Williams P. J. (1995) The effect of the nitrification inhibitor dicyandiamide on nitrate leaching and ammonia volatilization: A UK nitrate sensitive areas perspective. *Journal of Environmental Management* 45: 263-272.

de Klein C. A. M., Smith L. C. and Monaghan R. M. (2006) Restricted autumn grazing to reduce nitrous oxide emissions from dairy pastures in Southland, New Zealand. *Agriculture, Ecosystems and Environment* 112: 192-199.

de Klein C. A. M., van Logtestijn R. S. P., van de Meer H. G. and Geurink J. H. (1996) Nitrogen loss due to denitrification from cattle slurry injected into grassland soil with and without nitrification inhibitor. *Plant and Soil* 183: 161–170.

Di H. J., Cameron K. C., Silva R. G., Russell J. M. and Barnett J. (2002) A lysimeter study of the fate 15N-labelled nitrogen in cow urine with and without farm dairy effluent in a grazed dairy pasture soil under flood irrigation. *New Zealand Journal of Agricultural Research* 45: 235–244.

Dobbie K., McTaggart I. and Smith K. (1999) Nitrous oxide emissions from intensive agricultural systems: variations between crops and seasons, key driving variables, and mean emission factors. *Journal of Geophysical Research* 104: 26891-26899.

Ellis S., Yamulki S., Dixon E., Harrison R. and Jarvis S. (1998) Denitrification and N_2O emissions from a UK pasture soil following the early spring application of cattle slurry and mineral fertiliser. *Plant and Soil* 202: 15-25.

Flessa H. and Beese F. (2000) Laboratory estimates of trace gas emissions following surface application and injection of cattle slurry. *Journal of Environmental Quality* 29: 262-268.

Jarecki M. K, Parkin T. B., Chan A. S. K., Hatfield J. L. and Jones R. (2008) Greenhouse gas emissions from two soils receiving nitrogen fertilizer and swine manure slurry. *Journal of Environmental Quality* 37: 1432-1438.

Kirchmann H. and Lundvall A. (1998) Treatment of solid animal manures: identification of low NH_3 emission practices. *Nutrient Cycling in Agroecosystems* 51: 65-71.

Li J., Shi Y., Luo J., Houlbrooke D., Ledgard S., Ghani A. and Lindsey S. (2014) Effects of form of effluent, season and urease inhibitor on ammonia volatilization from dairy farm effluent applied to pasture. *Journal of Soils and Sediments* 14: 1341-1349.

Luo J., Ledgard S. and Lindsey S. (2008) A test of a winter farm management option for mitigating nitrous oxide emissions from a dairy farm. *Soil Use and Management* 24: 121-130.

Malgeryd J. (1998) Technical measures to reduce ammonia losses after spreading of animal manure. *Nutrient Cycling in Agroecosystems* 51: 51-57.

Merino P., Menéndez S., Pinto M., González-Murua C. and Estavillo J. M. (2005) 3, 4-Dimethylpyrazole phosphate reduces nitrous oxide emissions from grassland after slurry application. *Soil Use and Management*. 21: 53-57.

Misselbrook T. H., Nicholson F. A. and Chambers B. J. (2005) Predicting ammonia losses following the application of livestock manure to land. *Bioresource Technology* 96: 159-168.

Misselbrook T. H., Smith K. A., Johnson R. A. and Pain B. F. (2002) Structures and environment: slurry application techniques to reduce ammonia emissions: results of some UK field-scale experiments. *Biosystems Engineering* 81: 313-321.

Mkhabela M., Madani A., Gordon R., Burton D., Cudmore D., Elmi A., and Hart W. (2008) Gaseous and leaching nitrogen losses from no-tillage and conventional tillage systems following surface application of cattle manure. *Soil and Tillage Research* 98: 187-199.

Pain B. F., Phillips V. R., Clarkson C. R. and Karenbeck J. V. (1989) Loss of nitrogen through ammonia volatilization during and following the application of pig or cattle slurry to grassland. *Journal of Science Food and Agriculture* 47: 1–12.

Rochette P., Angers D. A., Chantigny M. H., Gagnon B. and Bertrand N. (2008) N_2O fluxes in soils of contrasting textures fertilized with liquid and solid dairy cattle manures. *Canadian Journal of Soil Science* 88: 175-187.

Rodhe L., Pell M. and Yamulki S. (2006) Nitrous oxide, methane and ammonia emissions following slurry spreading on grassland. Soil Use and Management 22: 229-237.

Rotz C. (2004) Management to reduce nitrogen losses in animal production. *Journal of Animal Science* 82: E119-E137.

Sommer S. G. and Hutchings N. (1995) Techniques and strategies for the reduction of ammonia emission from agriculture. *Water, Air and Soil Pollution* 85: 237–248.

Sommer S. G. and Olesen J. E. (1991) Effects of Dry Matter Content and Temperature on Ammonia Loss from Surface-Applied Cattle Slurry. *Journal of Environmental Quality* 20: 679-683.

Sommer S. G., Sherlock R. R. and Khan R. Z. (1996) Nitrous oxide and methane emissions from pig slurry amended soils. *Soil Biology and Biochemistry* 28: 1541-1544.

Sommer S. G., Zhang G. Q., Bannink A., Chadwick D., Misselbrook T., Harrison R., Hutchings N. J., Menzi H., Monteny G. J., Ni J. Q., Oenema O. and Webb J. (2006) Algorithms determining ammonia emission from buildings housing cattle and pigs and from manure stores. *Advances in Agronomy* 89: 261-335.

Stoeven K., Kohrs K. and Schnug E. (2002) Relation between N fertilization and N_2O release in different crop rotations. *Phyton (Austria)* 42: 191-200.

Thompson R. B., Pain B. F. and Rees Y. J. (1990) Ammonia volatilization from cattle slurry following surface application to grassland. *Plant and Soil* 125: 119–128.

Thomsen I. K., Pedersen A. R., Nyord T. and Petersen S. O. (2010) Effects of slurry pre-treatment and application technique on short-term N_2O emissions as determined by a new non-linear approach. *Agriculture, Ecosystems and Environment* 136: 227-235.

Thorman R., Chadwick D., Harrison R., Boyles L. and Matthews R. (2007) The effect on N_2O emissions of storage conditions and rapid incorporation of pig and cattle farmyard manure into tillage land. *Biosystems Engineering* 97: 501-511.

Vallejo A., García-Torres L., Díez J., Arce A. and López-Fernández S. (2005) Comparison of N losses (NO_3^-, N_2O, NO) from surface applied, injected or amended (DCD) pig slurry of an irrigated soil in a Mediterranean climate. *Plant and Soil* 272: 313-325.

van Groenigen J. W., Kasper G. J., Velthof G. L., van den Pol-van Dasselaar A. and Kuikman P. J. (2004) Nitrous oxide emissions from silage maize fields under different mineral nitrogen fertilizer and slurry applications. *Plant and Soil* 263: 101-111.

Watson C. J. (2000) Urease activity and inhibition Principles and practice. Proceedings-International Fertiliser Society York UK 40 pp.

Webb J., and Misselbrook T. H. (2004) A mass-flow model of ammonia emissions from UK livestock production. *Atmospheric Environment* 38: 2163-2176.

Webb J., Pain B., Bittman S. and Morgan J. (2010) The impacts of manure application methods on emissions of ammonia, nitrous oxide and on crop response - A review. *Agriculture, Ecosystems and Environment* 137: 39-46.

Webb J., Sommer S., Kupper T., Groenestein K., Hutchings N., Eurich-Menden B., Rodhe L., Misselbrook T. and Amon B. (2012) Emissions of ammonia, nitrous oxide and methane during the management of solid manures. In *Agroecology and Strategies for Climate Change*. Ed. E Lichtfouse. pp 67-107. Springer Netherlands.

Wulf S., Maeting M. and Clemens J. (2002) Application technique and slurry co-fermentation Effects on ammonia, nitrous oxide, and methane emissions after spreading. *Journal of Environmental Quality* 31: 1795-1801.

In: Greenhouse Gas Emissions ... ISBN: 978-1-53611-100-2
Editors: Jiafa Luo and Yan Li

Chapter 5

DIETARY MANIPULATION AS A TOOL FOR MITIGATING NITROUS OXIDE EMISSIONS

Sheree Balvert* and Jiafa Luo
AgResearch, Hamilton, New Zealand

5.1. INTRODUCTION

Currently, for every 100 units of nitrogen (N) used in agriculture, only about 15 are consumed as crop, dairy or meat products (Steinfeld et al. 2006; Robertson and Vitousek 2009). This points to very low N use efficiency in most agricultural systems. In the dairy industry, for example, despite improved genetic potential of cows with high nutrient utilisation, the increased external input of feed concentrates and use of fertiliser N have decreased N use efficiency in these systems (Huhtanen et al. 2008). Increasing N use efficiency is a key strategy by which the increasing food demand might be met without a corresponding increase in N losses and nitrous oxide (N_2O) emissions.

This chapter focuses on the use of dietary intervention or manipulation by incorporating diverse forages or crops into animal based agriculture as an approach for increasing N use efficiency. Dietary intervention is primarily aimed at either reducing the concentration and amount of N in animal excreta,

*Corresponding author: Sheree Balvert, AgResearch, 10 Bisley Road, Private Bag 3123, Hamilton 3214, New Zealand. Email: sheree.balvert@agresearch.co.nz.

or changing the partitioning of N in urine and dung, while preferably having no detrimental effect on productivity. This may include, but is not limited to, using low N feed supplements to reduce the amount of N consumed and hence lower the amount of surplus N being excreted, or specific dietary supplementation to encourage the partitioning of N toward dung rather than urine. Additionally, forages may be selected for their potential to inhibit soil N transformations by either the root exudation or urine-excretion of plant secondary metabolites. Dietary intervention may not be feasible in many low input, pasture based agricultural systems, however, in more intensive systems such as the New Zealand dairy industry, particularly with the increasing use of animal feed pads and housing practices, supplements are commonly imported and so dietary intervention is more practical.

5.2. Urine Patches and Nitrous Oxide

In order to understand the initiative to reduce excreta N, and urinary N in particular, it is necessary to understand the N dynamics of the urine patch in grazing systems and how they relate to N_2O emissions. Grazing ruminants are poor utilisers of dietary N, with only 10–30% of the N consumed being retained in body tissues and in products. The remainder is excreted in urine or dung (e.g., Selbie et al. 2015). This is generally because plants require more N to grow than ruminants require for protein synthesis (Haynes and Williams 1993) and so excess N is consumed by the ruminant. There have been several studies that have demonstrated a strong linear relationship between N intake and N excretion (e.g., Castillo et al., 2000; Kebreab et al., 2001; Mulligan et al. 2004; Huhtanen et al., 2008) and in New Zealand about 80% of the total N_2O emissions come from urine and dung deposition.

Urine deposited onto grazed pastures is considered a hotspot for N_2O production as urine provides high localised concentrations of available N and carbon in soils. In New Zealand, the average urine N loading rate is about 600 kg N ha^{-1} for dairy cattle, and about 1000 kg N ha^{-1} for sheep (Selbie et al. 2015). Additionally, other abiotic conditions found in the urine patch, in particular soil water filled pore space, encourage anaerobic denitrification which is the main N_2O forming pathway.

5.3. Reducing Dietary N

Due to the strong relationship between N intake and N excretion, one of the simpler ways to reduce N excreted (and hence N available as a substrate for N_2O production) is to reduce the N consumed in the diet through the use of diverse feeds. As N in pasture is generally in excess of animal requirements, reducing N intake should presumably not reduce production. Care however must be taken to ensure dietary N/protein is not reduced to a point where it is no longer sufficient to meet the animal's metabolic requirements.

Many studies have shown that reducing N content in the diet through low N supplements reduces total and urinary N excreted in ruminants (e.g., Edwards et al. 2014; Huhtanen et al. 2008; Kebreab et al. 2001; Misselbrook et al. 2005; Mulligan et al. 2004 (and references therein); Nielsen et al. 2003; Steinshamn et al. 2006). In a New Zealand example, Luo et al. (2008) in a study on a pasture based dairy system, showed that using low N feed supplements such as maize to reduce the amount of N consumed, significantly reduced the amount of surplus N being excreted and showed an increase in milk production per unit of N and thus overall N use efficiency. Additionally, Burke et al. (2008) carried out a study on an Irish dairy system, which showed a reduction in both urinary and total N excretion, and no effect on production, when supplementing the diet with low N citrus pulp.

5.4. N Partitioning – Protein Vs Excreta

The low level of utilisation of dietary N is largely a result of insufficient carbohydrate-derived energy for conversion of the feed N to microbial N in the rumen. The surplus N is then lost through the rumen wall as ammonia and subsequently excreted in urine following hepatic conversion to urea N (Miller et al. 2001). It therefore suggests that increasing dietary soluble carbohydrate content may increase utilisation of feed N in the rumen and thus reduce N excretion.

Miller et al. (2001) and Moorby et al. (2006) showed that the addition of 'high sugar grasses', or grass varieties with a high content of water soluble carbohydrate, to the diet increased milk production in dairy cows through more efficient utilisation of the feed N in the rumen. They also measured a reduction in total excreted N and urinary N. Additionally, Lee et al. (2002) found reduced ammonia-N concentrations in the rumen of steers fed high sugar grasses compared with steers offered the control perennial ryegrass variety, which translated through to reduced urine N excretion. These results are supported by others (e.g., Castillo et al. 2000; Broderick 2003; Steinshamn et al. 2006) who have reported improved N utilisation and reduced excretal N following supplementation of high energy feeds to cattle.

5.5. N Partitioning – Dung vs Urine

Although the relationship between N intake and N excreted is linear for both dung and urine, the line is much steeper for urine than for dung (e.g., Dijkstra et al. 2013), indicating that urine is the main removal pathway for excess dietary N. N_2O emissions from dung deposits are considerably lower than from urine patches as the N transformation processes are much slower in dung, and it has been found that the emission factor (percentage of applied N emitted as N_2O-N) for dung is significantly lower than that of urinary N (van der Weerden et al. 2011). It is therefore likely that encouraging excess dietary N to be excreted in dung rather than urine would result in an overall reduction of N_2O emissions.

Supplementation of diet with condensed tannins, for example, has been shown to result in a greater partitioning of excreta N in dung rather than urine. Condensed tannins bind with proteins in the rumen, slowing protein degradation and leading to decreased ruminal NH_3 concentration and subsequently decreased urinary N excretion, without affecting production (Carulla et al. 2005; Misselbrook et al. 2005).

5.6. Plant Secondary Metabolites

There is extensive evidence to show that plant tissues contain a wide variety of chemical compounds which are not involved in primary metabolism. These compounds, termed ‘secondary metabolites’, vary according to family and species and play a variety of roles in protecting plants from a range of stresses (e.g., predation and competition, Bennett and Wallsgrove 1994; Erickson et al. 2000) including through suppression or inhibition of soil N transformation processes. Inhibitors of soil N processes, such as urease and nitrification inhibitors, have the potential to reduce N_2O emissions from animal excreta deposited to land (Dell et al. 2014; Li et al. 2014). Urease inhibitors retard the hydrolysis of urea to NH_4^+ whereas nitrification inhibitors retard the conversion of NH_4^+ to NO_3^-. These both prolong the length of time taken for applied N (e.g., excreta or fertiliser N) to be converted to NO_3^- thereby allowing for more uptake by plants and improving N use efficiency (Amberger 1989). This results in less available NO_3^- for both direct N_2O emissions and indirect emissions following leaching.

The suppression of nitrification has been observed to occur naturally in some ecosystems where certain plant species release organic molecules from their roots that suppress the function and growth of nitrifying bacteria (Subbarao et al. 2007). This process is termed biological nitrification inhibition (BNI). It has been hypothesised that BNI has evolved as part of some plants' adaptation mechanisms to conserve and use N efficiently in systems that are naturally limited in mineral N (Al-Ansari and Abdulkareem 2014; Subbarao et al. 2007) and appears to be stimulated by high NH_4^+ concentration in the soil (Subbarao et al. 2015). BNI has the potential to improve agronomic nutrient use efficiency and, as a result, reduce both leached and gaseous N loss from agricultural systems. A field study carried out over two years in China, showed higher vegetable yield and nutrient use efficiency for crops co-planted with known biological nitrification inhibiting plants, as well as a reduction in N_2O emissions (Zhang et al. 2015). Several BNIs have been isolated from root exudates and plant tissues and BNIs that belong to many different chemical classes have been identified (Gopalakrishnan et al. 2007; Subbarao et al. 2013).

Additionally, residues of plant tissues are also known to release allelo-chemicals that inhibit nitrification in soil (Bremner and McCarty 1993). As these chemicals are released during destruction or decomposition of the plant, it is thought they are a defence mechanism against herbivores, pests or pathogens, but also inhibit nitrification (Bending and Lincoln 2000; Bennett and Wallsgrove 1994). Some examples of plants where tissue residues or extracts of tissues inhibit nitrification include neem, karanj, plantain, sorghum and brassica species (Luo unpublished data; Prasad and Power 1995; Zhang et al. 2015).

Incorporating plant tissues into soil may inhibit nitrification in cropping systems, however, this approach is less suitable for grazing systems where pastures are generally only renewed every 5-10 years so there is little opportunity to incorporate plant material. Also, N substrate supplied in the excreta of grazing animals is the main driver of nitrification in these systems. Therefore, rather than incorporating plant residues, it may be possible to deliver BNI compounds to the soil through the grazing animal. Grazers eliminate unprocessed plant material, and presumably secondary metabolites, in excreta (dung and urine). One advantage of this approach is that the inhibitor is delivered with the main N source for N_2O production. For example, Ledgard et al. (2008) and Welten et al. (2013) showed that urine patches could be individually targeted with a synthetic nitrification inhibitor (in this instance dicyandiamide) by orally administering the inhibitor to grazing animals that was subsequently excreted in the urine. It is possible that this method could also be used for administering plant derived inhibitors to urine patches. Luo et al. (2015) recently reported reduced N_2O emissions from urine patches derived from animals fed a brassica crop compared to urine from animals fed on perennial pasture. The N content of urine from brassica fed animals is lower than that of ryegrass fed animals and so lower N_2O emissions would be expected, however, they also reported reduced emission factors which accounts for the differences in N loading. One possible explanation for this reduction is that brassica-derived compounds may be delivered to the soil in the urine and nitrification may subsequently be inhibited in the urine patch. There is an array of potential BNIs that could possibly be utilised for reducing N_2O production in agricultural systems. In New Zealand systems, fodder crops that are known to have BNI potential include: leafy turnips, bulb turnips, swedes, kale, forage rape and plantain (de Ruiter et al. 2009). However, there is little research that has examined the potential for these feeds to reduce nitrification and N_2O emissions in practice.

CONCLUSION

This chapter outlines the potential for using dietary manipulation as a tool for mitigating N_2O emissions, and indeed improving N use efficiency in general, from intensive animal agricultural systems.

Most of the approaches are aimed at reducing N excreted in the urine patch and associated N_2O formed under patch conditions. A number of studies have already been carried out in this area and much is known about which feed characteristics can be used to achieve this. It is however, important to ensure that animal metabolic requirements are always met.

Other approaches focus on inhibiting soil N transformation processes through the use of plant secondary metabolites. Although there are studies that confirm the presence and potential of such compounds, their effectiveness via a feed manipulation route requires further testing.

ACKNOWLEDGMENTS

The authors acknowledge funding from the New Zealand Agricultural Greenhouse Gas Research Centre. The authors also acknowledge guidance from Louis Schipper from the University of Waikato.

REFERENCES

Al-Ansari A. M. S. and Abdulkareem M. A. (2014) Some plant extracts retarde nitrification in soil. *Acta Agriculturae Slovenica* 103: 5-13.

Amberger A. (1989) Research on dicyandiamide as a nitrification inhibitor and future outlook. *Communications in Soil Science and Plant Analysis* 20: 1933-1955.

Bending G. D. and Lincoln S. D. (2000) Inhibition of soil nitrifying bacteria communities and their activities by glucosinolate hydrolysis products. *Soil Biology and Biochemistry* 32: 1261-1269.

Bennett R. N. and Wallsgrove R. M. (1994) Tansley Review No. 72. Secondary metabolites in plant defence mechanisms. *New Phytologist* 127: 617-633.

Bremner J. M. and McCarty G. W. (1993) Inhibition of nitrification in soil by allelochemicals derived from plants and plant residues. *Soil Biochemistry* 8: 181-218).

Broderick G. A. (2003) Effects of Varying Dietary Protein and Energy Levels on the Production of Lactating Dairy Cows1. *Journal of Dairy Science* 86: 1370-1381.

Burke F., O'Donovan M. A., Murphy J. J., O'Mara F. P. and Mulligan F. J. (2008) Effect of pasture allowance and supplementation with maize silage and concentrates differing in crude protein concentration on milk production and nitrogen excretion by dairy cows. *Livestock Science* 114: 325-335.

Carulla J. E., Kreuzer M., Machmüller A. and Hess, H. D. (2005) Supplementation of Acacia mearnsii tannins decreases methanogenesis and urinary nitrogen in forage-fed sheep. *Australian Journal of Agricultural Research* 56: 961-970.

Castillo A. R., Kebreab E., Beever D. E. and France J. (2000) A review of efficiency of nitrogen utilisation in lactating dairy cows and its relationship to environmental pollution. *Journal of Animal and Feed Sciences* 9: 1-32.

de Ruiter J., Wilson D., Maley S., Fletcher A., Fraser T., Scott W., Berryman S., Dumbleton A. and Nichol W. (2009) *Management practices for forage brassicas*. New Zealand. pp 70.

Dell C. J., Han K., Bryant R. B. and Schmidt J. P. (2014) Nitrous oxide emissions with enhanced efficiency nitrogen fertilizers in a rainfed system. *Agronomy Journal* 106: 723-731.

Dijkstra J., Oenema O., van Groenigen J. W., Spek J. W., van Vuuren A. M. and Bannink A. (2013) Diet effects on urine composition of cattle and N_2O emissions. *Animal* 7(s2): 292-302.

Edwards G. C., De Ruiter J. M., Dalley D. E., Pinxterhuis J. B., Cameron K. C., Bryant R. H., Di H. J., Malcolm B. J. and Chapman D.F. (2014) Urinary nitrogen concentration of cows grazing fodder beet, kale and kale-oat forage systems in winter. *Australian Dairy Science Symposium:* 144-147.

Erickson A. J., Ramsewak R. S., Smucke, A. J. and Nair M. G. (2000) Nitrification Inhibitors from the roots of Leucaena leucocephala. *Journal of Agricultural and Food Chemistry* 48: 6174-6177.

Gopalakrishnan S., Subbarao G. V., Nakahara K., Yoshihashi T., Ito O., Maeda I., Ono H. and Yoshida M. (2007) Nitrification inhibitors from the root tissues of Brachiaria humidicola, a tropical grass. *Journal of Agricultural and Food Chemistry 55*: 1385-1388.

Haynes R. J. and Williams P. H. (1993) Nutrient cycling and soil fertility in the grazed pasture ecosystem. *Advances in Agronomy* 49: 119-199.

Huhtanen P., Nousiainen J. I., Rinne M., Kytölä K. and Khalili H. (2008) Utilization and partition of dietary nitrogen in dairy cows fed grass silage-based diets. *Journal of Dairy Science* 91: 3589-3599.

Kebreab E., France J., Beever D. E. and Castillo A. R. (2001) Nitrogen pollution by dairy cows and its mitigation by dietary manipulation. *Nutrient Cycling in Agroecosystems* 60: 275-285.

Ledgard S. F., Menneer J. C., Dexter M. M., Kear M. J., Lindsey S., Peters J. S. and Pacheco D. (2008) A novel concept to reduce nitrogen losses from grazed pastures by administering soil nitrogen process inhibitors to ruminant animals: A study with sheep. *Agriculture, Ecosystems and Environment* 125: 148-158.

Lee M. R. F., Harris L. J., Moorby J. M., Humphreys M. O., Theodorou M. K., MacRae J. C. and Scollan N. D. (2002) Rumen metabolism and nitrogen flow to the small intestine in steers offered Lolium perenne containing different levels of water-soluble carbohydrate. *Animal Science* 74: 587-596.

Li J., Shi Y., Luo J., Zaman M., Houlbrooke D., Ding W., Ledgard S. and Ghani A. (2014) Use of nitrogen process inhibitors for reducing gaseous nitrogen losses from land-applied farm effluents. *Biology and Fertility of Soils* 50: 133-145.

Luo J., Ledgard S. F., de Klein C. A. M., Lindsey S.B. and Kear M. (2008) Effects of dairy farming intensification on nitrous oxide emissions. *Plant and Soil* 309: 227-237.

Luo J., Sun X. Z., Pacheco D., Ledgard S., Lindsey S., Hoogendoorn C.J., Wise B. and Watkins N. (2015). Nitrous oxide emission factors for urine and dung from sheep fed either fresh forage rape (Brassica napus L.) or fresh perennial rygrass (Lolium perenne L.). *Animal* 9: 534-543.

Miller L. A., Moorby J. M., Davies D. R., Humphreys M. O., Scollan N. D., MacRae J. C. and Theodorou M. K. (2001) Increased concentration of water-soluble carbohydrate in perennial ryegrass (Lolium perenne L.): milk production from late-lactation dairy cows. *Grass and Forage Science* 56: 383-394.

Misselbrook T. H., Powell J. M., Broderick G. A. and Grabber J. H. (2005) Dietary manipulation in dairy cattle: Laboratory experiments to assess the influence on ammonia emissions. *Journal of Dairy Science* 88: 1765-1777.

Moorby J. M., Evans R. T., Scollan N. D., MacRae J. C. and Theodorou M. K. (2006) Increased concentration of water-soluble carbohydrate in perennial ryegrass (Lolium perenne L.). Evaluation in dairy cows in early lactation. *Grass and Forage Science* 61: 52-59.

Mulligan F. J., Dillon P., Callan J. J., Rath M. and O'Mara F. P. (2004) Supplementary concentrate type affects nitrogen excretion of grazing dairy cows. *Journal of Dairy Science* 87: 3451-3460.

Nielsen N. M., Kristensen T., Nørgaard P. and Hansen H. (2003) The effect of low protein supplementation to dairy cows grazing clover grass during half of the day. *Livestock Production Science* 81: 293-306.

Prasad R. and Power J. F. (1995) Nitrification inhibitors for agriculture, health, and the environment. *Advances in Agronomy* 54: 233-281.

Robertson G. P. and Vitousek P. M. (2009) Nitrogen in agriculture: Balancing the cost of an essential resource. *Annual Review of Environment and Resources* 34: 97-125.

Selbie D. R., Buckthought L. E. and Shepherd M. A. (2015) The challenge of the urine patch for managing nitrogen in grazed pasture systems. *Advances in Agronomy* 129: 229-292.

Steinfeld H., Gerber P., Wassenaar T., Castel V., Rosales M. and de Haan C. (2006) Livestock's long shadow: Environmental issues and options. *Renewable Resources Journal* 24: 15-17.

Steinshamn H., Höglind M., Garmo T. H., Thuen E. and Brenøe U. T. (2006) Feed nitrogen conversion in lactating dairy cows on pasture as affected by concentrate supplementation. *Animal Feed Science and Technology* 131: 25-41.

Subbarao G. V., Rao I. M., Nakahara K., Ando Y., Sahrawat K. L., Tesfamariam T., Lata J. C., Boudsocq S., Miles J. W., Ishitani M. and Peters M. (2013) Nitrogen management in grasslands and forage-based production systems - Role of biological nitrification inhibition (BNI). *Tropical Grasslands - Forajes Tropicales* 1: 168-174.

Subbarao G. V., Rondon M., Ito O., Ishikawa T., Rao I. M., Nakahara K., Lascano C. and Berry W. (2007) Biological nitrification inhibition - is it a widespread phenomenon? *Plant and Soil* 294: 5-18.

Subbarao G. V., Yoshihashi T., Worthington M., Nakahara K., Ando Y., Sahrawat, K. L., Rao I. M., Lata J. C., Kishii M. and Braun H.J. (2015) Suppression of soil nitrification by plants. *Plant Science* 233: 155-164.

van der Weerden T. J., Luo J., de Klein C. A. M., Hoogendoorn C. J., Littlejohn R. P. and Rys G. J. (2011) Disaggregating nitrous oxide emission factors for ruminant urine and dung deposited onto pastoral soils. *Agriculture, Ecosystems and Environment* 141: 426-436.

Welten B. G., Ledgard S. F., Schipper L. A. and Judge A. A. (2013) Effect of amending cattle urine with dicyandiamide on soil nitrogen dynamics and leaching of urinary-nitrogen. *Agriculture, Ecosystems and Environment* 167: 12-22.

Zhang M., Fan C. H., Li Q. L., Li B., Zhu Y. Y. and Xiong Z. Q. (2015) A 2-yr field assessment of the effects of chemical and biological nitrification inhibitors on nitrous oxide emissions and nitrogen use efficiency in an intensively managed vegetable cropping system. *Agriculture, Ecosystems and Environment* 201: 43-50.

In: Greenhouse Gas Emissions ...
Editors: Jiafa Luo and Yan Li
ISBN: 978-1-53611-100-2

Chapter 6

Optimal Cattle Manure Application Rates to Maximise Crop Yield and Minimise Risk of Nitrogen Loss to the Environment

Yongping Jing[1,2,3,*], Yan Li[1,2,3], Yingpeng Zhang[1,2], Jiafa Luo[1], Luji Bo[1,2,3], Ming Sun[1] and Ziwen Zhong[1]

[1]Institute of Agricultural Resources and Environment, Shandong Academy of Agricultural Sciences, Jinan, China
[2]Key Laboratory of Agro-Environment in Huang-Huai-Hai Plain, Ministry of Agriculture, Jinan, China
[3]Shandong Provincial Key Laboratory of Plant Nutrition and Fertilizer, Jinan, China

6.1. Introduction

With rapid livestock intensification in China, application of large amounts of livestock manure can be an important part of agricultural non-point source pollution (Zhang et al. 2007). However, as livestock manure contains nutrients for plant growth, use of the manure as a fertilizer on farmland can be an

[*] Corresponding author: Dr Yongping Jing, Institute of Agricultural Resources and Environment, Shandong Academy of Agricultural Sciences, Jinan 250100, China. Email: jyp82916@163.com.

efficient way to use this resource. Many studies have shown that application of manure can increase availability of soil nutrients and improve soil fertility and organic matter content (Bolan 2003; Yu et al. 2012). However, an increasing number of studies have also shown that the excessive application of animal manure, such as cattle manure, can result in nitrate (NO_3^-) accumulation in soil, which could lead to NO_3^- pollution to surface water and groundwater. Yuan et al. (2000) found that the surface runoff from continuous annual application of 20 tons/ha/year of poultry manure has imposed a serious threat to surface water quality. Furthermore, Yao et al. (2005) and Liu et al. (2008) found that high application rates of chicken manure could make NO_3^--N concentration in soil layers above 4 m increase to 40–50 mg/kg. The research on the Taihu basin by Lin et al. (2003) showed that the contribution of nitrogen (N) to water pollution, 80% of N in field runoff and leachate were derived from N mineralized from organic manure. N leaches into the groundwater through the action of rainfall or irrigation when fertilizer application rate exceeds soil denitrification capacity (Salvador et al. 2009). Therefore, determining an optimal application rate for animal manure to increase production while avoiding adverse environmental effects is required.

Wheat–maize rotation, as one of the most important cropping systems in the North China Plain, is characterized by high nutrient input and output. Detailed studies of NO_3^--N accumulation and migration in the soil profile in the wheat–maize rotation cropping system, when cattle manure is applied, would help in understanding the leaching risk of applied manure N, so that efficient application methods can be developed.

This chapter presents a study in which the effect of different cattle manure application rates on NO_3^--N migration and accumulation in soil is analysed. The effects on crop yield and manure N use efficiency are also presented. The aim of this study was to provide data for determination of optimal manure application rates in order to control N loss from farmland and maximize crop yields in the North China Plain. In addition, the data from this study will add to a database that provides a theoretical basis for formulation of a national standard for livestock manure loading rates for arable land in China.

6.2. Materials and Methods

6.2.1. Overview of the Experiment Site

The experiment was carried out in Wanglou village, Caoxian county of Shandong Province, China, 34°47′9.78″ N, 115°35′16.32″ E. The area is in the north subtropical region, with an annual mean temperature of 13.6–14.3 °C, annual mean sun hours of 2148 h, annual precipitation of 612–711 mm and a mean frost-free period of 212 days. The cropping system was an annual winter wheat–summer maize rotation. The soil at the study site is an alluvial soil, and the basic physical and chemical properties of the 0–20 cm layer were as follows: pH (H_2O, 1:5) of 8.56, organic matter content of 13.25 g/kg, total N of 1.33 g/kg, available P of 17.76 mg/kg, alkali-hydrolyzable N of 93.54 mg/kg and available K of 129.17 mg/kg.

6.2.2. Experimental Design

Winter wheat was sown on October 9th, 2014 and harvested on June 4th, 2015. Maize was sown on June 4th, 2015 and harvested on October 10th, 2015. The organic fertilizer used for the experiment was cattle manure from a nearby dairy farm. The feed for the cattle was mainly maize silage with alfalfa and wheat grain as supplements. The cattle manure was produced through a composting process after dry-and-wet separation. Manure nutrient contents are shown in Table 6.1. Six treatments with different manure application rates were applied. The manure application rates were 0, 5000, 30000, 45000, 60000 and 75000 kg/ha/year, corresponding to N rates of 0, 75, 450, 675, 900 and 1125 kg/ha/year. Half of the manure was applied to the wheat and the other half to the maize. Manure was applied at the surface at the time of sowing of the wheat or maize, and mixed well with the soil through mechanical rotary tillage at 20 cm depth. Seeds were sown after soil levelling and there was no fertilizer topdressing at later stages for either crop.

Table 6.1. Nutrient content of the cattle manure

N (%)	P_2O_5 (%)	K_2O (%)	Organic matter (%)	Solid content (%)
1.50	1.70	1.48	29.01	57.07

The wheat cultivar used was Jimai 22, which is locally prevalent, and mechanical row sowing of 25 cm was adopted with a seeding rate of 187 kg/ha. The maize cultivar used was Yuqingchu 23 and was sown using the dibbling method with an even row spacing of 60 cm and plant spacing of 25 cm. All treatments had three replicates and were randomly arranged in blocks with a plot size of 64 m^2. The management practices for the experimental plots were the same as those in local cropping fields. Total precipitation was 179 mm during the wheat growth season and 468 mm during the maize growth season (Figure 6.1).

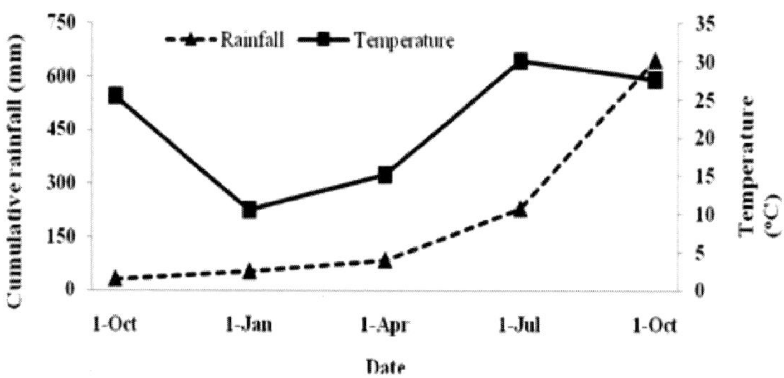

Figure 6.1. Precipitation and air temperature during the wheat and maize growing seasons.

6.2.3. Sample Collection

Soil samples were collected from each plot at the seedling, heading and maturing stages for both wheat and maize. Five layers of the soil profile were sampled with each layer 20 cm in depth. Three cores were taken and pooled from each plot randomly. The field-moist soils were sieved (<4 mm) soon after collection and then analysed for soil moisture and NO_3^--N. The soil bulk density of each layer was determined (Lu 1999). Accumulation of NO_3^- in the soil profile was determined based on the NO_3^--N content of each soil layer and the corresponding soil bulk density. At the maturing stage, wheat grains and straws and maize silage were sampled for determination of crop yields and N levels. Cattle manure N use efficiency was calculated as the percentage of applied cattle manure N which was taken up by crops.

When the wheat was harvested, samples were taken from a 5 m^2 area for yield measurement and simultaneous nutrient content determination in wheat grains and straws. When the maize silage was harvested, 10 strains of maize plants were selected randomly to determine the whole-maize-plant biomass and nutrient content and the whole plot was harvested for yield calculation.

6.2.4. Determination Methods

For determination of NO_3^--N and ammonium N (NH_4^+-N) in the soil, fresh soil was extracted with 2 M KCl. The extracts were filtered and the filtrates analysed using a FiaStar 5000 flow injection analyzer (Foss, Sweden). The total N of the plant was determined using the semi-micro Kjeldahl N method after digesting samples with concentrated sulfuric acid-hydrogen peroxide solution (Bao et al. 2000).

6.2.5. Calculation Methods

The amount of NO_3^--N accumulated in each soil layer was calculated based on the NO_3^--N content of each layer (20 cm) and the soil bulk density: $R_i = C \times D \times H \times A \times 10^{-3}$, where R_i is the NO_3^--N accumulation amount for each soil layer (kg/hm^2), C is the soil NO_3^--N content (mg/kg), D is soil bulk density (g/cm^3), H is the soil layer thickness (m) and A is land area (hm^2; 100 × 100 m).

The total NO_3^--N accumulation in a certain soil profile was calculated as the sum of the NO_3^--N accumulation in each soil layer of the soil profile (Wang et al. 2007).

Agronomic use efficiency of N fertilizer (kg grain/kg N) was calculated as (crop yield in the area with N application − crop yield in the area without N application) / N application rate.

The Nitrogen fertilizer use ratio (%) using the balancing method was calculated as crop N uptake / crop N input × 100 (Sobańtka et al. 2010).

All data were statistically analyzed using SPSS 16.0 software. The significance of variation was determined by using the least significant difference method after a single factor variance analysis, with $P < 0.05$ as the significance level.

6.3. RESULTS

6.3.1. Distribution of NO_3^--N in the Soil Profile during the Wheat Season

At the wheat seedling, heading and maturing stages, NO_3^--N concentration for soil layers from 0 to 100 cm increased with increasing cattle manure application rate (Figure 6.2). On the 59th day after manure application (wheat seedling stage), soil NO_3^--N contents in the top soil layers (0–40 cm) of the plots with the higher manure application rates were significantly higher than those with lower application rates ($P < 0.05$) (Figure 6.2A). In the 40–100 cm soil layers, NO_3^--N contents in the manure treated soils were higher than those in the control ($P < 0.05$), but differences in NO_3^--N content between manure application rates were not significant ($P > 0.05$). Total rainfall for those 59 days was 33 mm and temperature was below 5 °C (Figure 6.1). These data suggest that mineralization of manure organic N had started and some NO_3^--N had gradually moved to the deeper soil layers, despite the low rainfall and temperature for over that time.

On the 188th day after manure application (heading stage) NO_3^--N contents had further increased in the soil profile (Figure 6.2B) as a result of more organic manure N being mineralized, with more rainfall and higher temperature. In the 0–40 cm soil layers, the NO_3^--N content of the treatments with the two highest manure application rates were significantly higher than those with the two lowest rates ($P < 0.05$). Also, much higher soil NO_3^--N contents were found in the 40–100 cm soil layers in the treatments with the two highest manure application rates than in the treatments with lower application rates.

On the 237th day after manure application (maturing stage), differences in NO_3^--N contents in the deep layers of soil (60-100 cm) between manure application rates were significant ($P < 0.05$), with much higher NO_3^--N contents in the treatments with the two highest manure application rates (Figure 6.2C). The accumulation of NO_3^--N in the deep soil layers (out of the root zones) posed a great risk for leaching to groundwater.

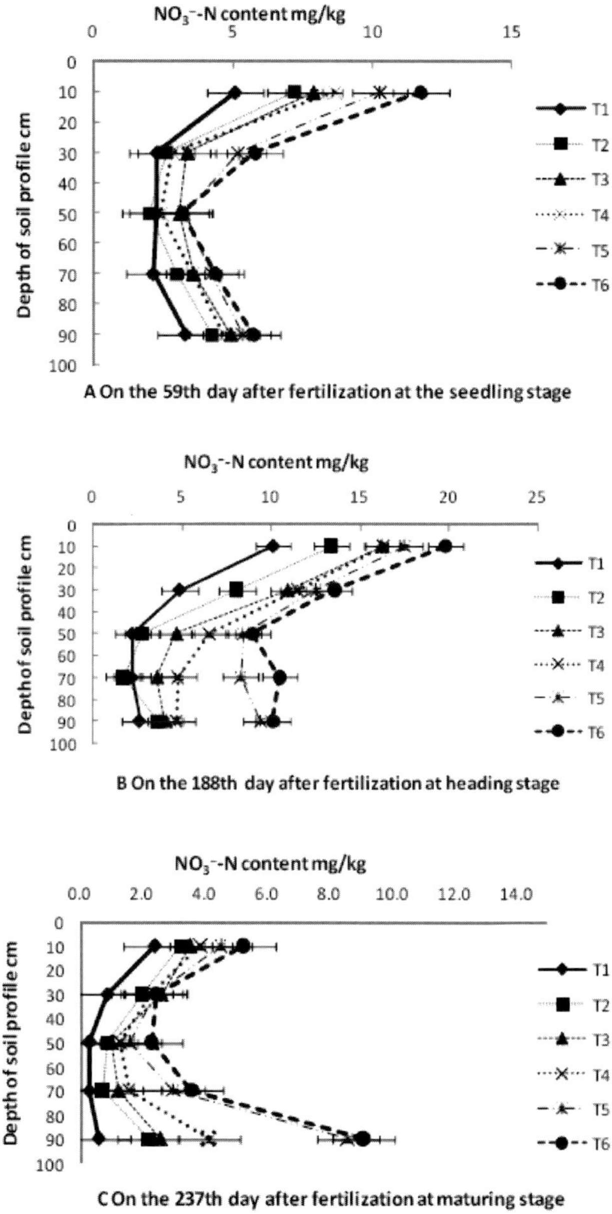

Figure 6.2. Spatiotemporal variations of soil NO_3^--N contents in different treatments of cattle manure during the maize growing season (T1 = control; T2 = 15000 kg/ha; T3 = 30000 kg/ha; T4 = 45000 kg/ha; T5 = 60000 kg/ha; and T6 = 75000 kg/ha).

6.3.2. Distribution of NO_3^--N in the Soil Profile during the Maize Season

The maize growth season coincided with the raining season, and a much higher amount of rainfall was recorded for the maize growth period (82 days) than for the wheat growth period (240 days) (Figure 6.1). Generally, higher NO_3^--N contents were observed in the soil profiles with higher manure application rates (Figure 6.3). On the 20th day after manure application (maize seedling stage), soil NO_3^--N contents in the 0–60 cm soil layers were not significantly different between the treatments ($P > 0.05$) (Figure 6.3A). However, the NO_3^--N contents in the 60–100 cm soil layers in the treatment with the highest manure application rate (75000 kg/ha) were significantly higher than those in all other treatments ($P < 0.05$).

On the 40th day after manure application (maize heading stage), soil NO_3^--N contents in the 0–60 cm soil layers in the treatment with the highest manure application rate were significantly higher than those at the other application rates ($P < 0.05$) (Figure 6.3B). On the 82nd day after manure application (maturing stage), soil NO_3^--N contents in the 0–40 cm soil layers of the 75000 kg/ha treatment were significantly higher than in the other treatments ($P < 0.05$) (Figure 6.3C). However, in the 40–60 soil layers, the NO_3^--N contents were not significantly different between the treatments.

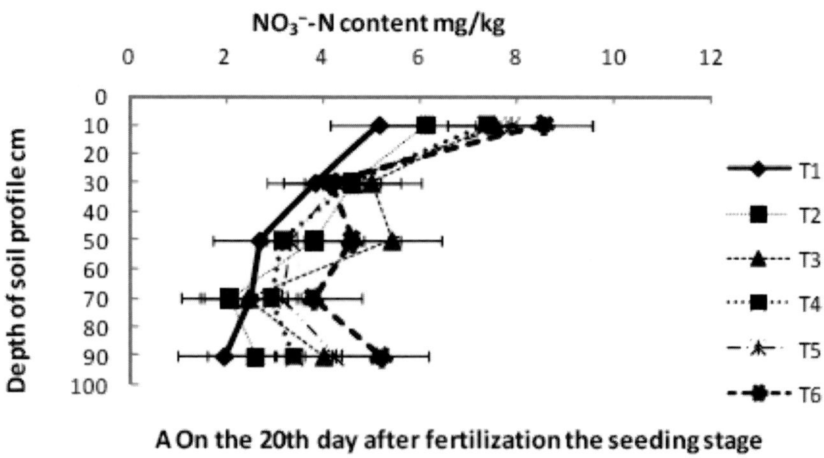

Figure 6.3. (Continued)

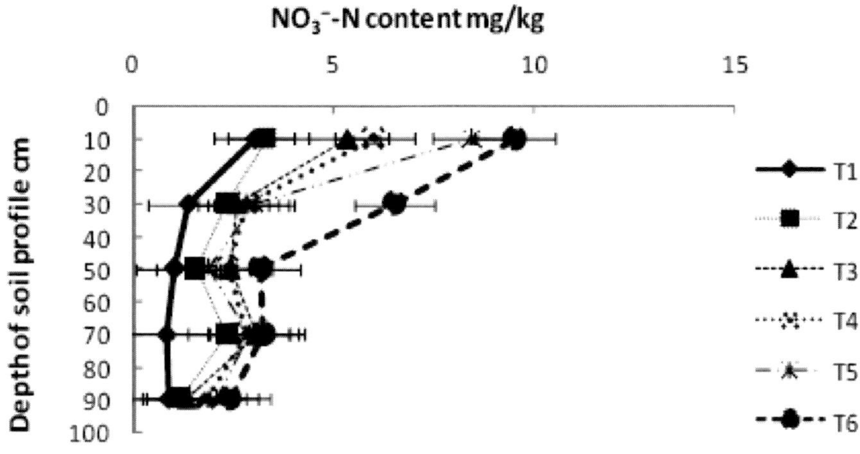

B On the 40th day after fertilization at jointing stage

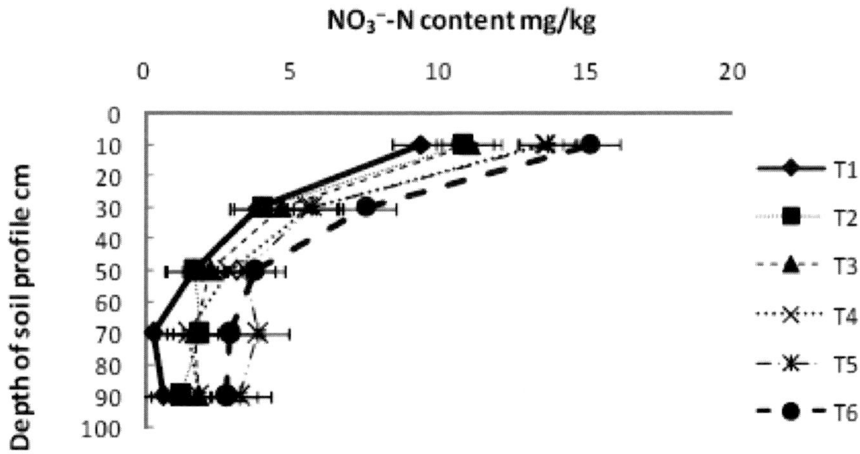

C On the 82nd day after fertilization at the harvesting stage

Figure 6.3. Spatiotemporal variations of soil NO_3^--N contents of different treatments of cattle manure during the maize growing season (T1 = control; T2 = 15000 kg/ha; T3 = 30000 kg/ha; T4 = 45000 kg/ha; T5 = 60000 kg/ha; and T6 = 75000 kg/ha).

Although higher NO_3^--N contents in the 60–100 cm soil layers in the 75000 kg/ha treatment were observed at the heading and maturing stages, the differences were not significant ($P > 0.05$). It is suggested that by these later stages of the maize growing period the NO_3^--N is likely to have been drained out of the soil profile (below 100 cm), without accumulation in the soil, as a result of continuous drainage due to the large amount of rainfall during the season.

6.3.3. Variation in Soil NO_3^--N Accumulation over Time

The NO_3^--N accumulation in the 0–100cm soil profile was influenced greatly by temperature, rainfall and plant uptake. The highest NO_3^--N accumulation occurred at the wheat heading stage for each treatment, with NO_3^--N accumulation of up to 176 kg/ha (Figure 6.4). Some of this accumulated NO_3^--N would be prone to leaching out of the soil profile when the rainfall season commenced. There was a trend for increasing amounts of NO_3^--N to accumulate in the soil with increasing manure application rate, with significantly higher amounts of NO_3^--N generally being observed in the soil from the two highest manure application rates.

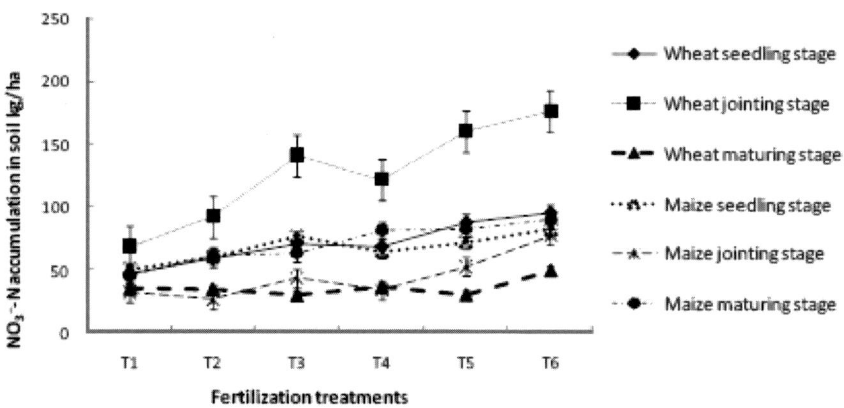

Figure 6.4. Changing pattern of NO_3^--N accumulation in 0–100 cm soil over time after cattle manure addition (T1 = control; T2 = 15000 kg/ha; T3 = 30000 kg/ha; T4 = 45000 kg/ha; T5 = 60000 kg/ha; and T6 = 75000 kg/ha).

6.3.4. Effect of Manure Application Rate on Crop Yield and N Use Efficiency

Wheat yield increased with increasing manure application rate up to 45000 kg/ha, but decreased when the application rates were higher (Table 6.2).

The relationship between the wheat yield (kg/ha) and manure application rate (kg/ha) can be best described by a quadratic function: $y = -4\times10^{-5}x^2 + 0.1232x + 279.42$, $R^2 = 0.92$. Maize silage yield increased with increasing manure application rate. The yield (t/ha) and manure rate (kg/ha) showed a significant and linear relationship ($y = 0.0006x + 3.3249$, $R^2 = 0.8827$). The agronomic efficiency of applied manure N was highest when the manure application rate was 15000 kg/ha (Table 6.2); however, the crop yields were lower at this application rate. The corresponding cattle manure application rates for the highest wheat and maize silage yields were 45000 and 75000 kg/ha, respectively, but the maize silage yield showed no significant increase when the application rate was increased from 45000 to 75000 kg/ha. Therefore, the optimum cattle manure application rate would be 45000 kg/ha in the wheat-maize rotation cropping system.

Table 6.2. Wheat and maize silage yields and N efficiency of treatments with different application rates

	Wheat			Maize		
Total manure rate (kg/ha)	Yield (kg/ha)	N agronomic efficiency (kg wheat/kg applied N)	N use efficiency (%)	Yield (ton/ha)	N agronomic efficiency (kg maize/kg applied N)	N use efficiency (%)
0	4134 c			45b		
15000	5034b	8.0	125	59ab	12.0	120
30000	5424ab	5.7	76	60ab	6.5	71
45000	5704a	4.7	52	65a	5.8	50
60000	5136b	2.2	33	68a	5.1	43
75000	5072b	1.7	25	71a	4.5	32

Note: In each column of data, different letters represent a significant difference among treatments, $P < 0.05$; N agronomic efficiency (kg yield/kg N) = (crop yield in N application area − crop yield control) /Manure N application rate. N use efficiency was calculated as the percentage of applied cattle manure N which was uptaken by crops.

The N uptake by wheat and maize increased with increasing manure application rate up to 45000 and 60000 kg/ha, respectively. However, the N uptake decreased when the manure rates were further increased (Table 6.2). Manure N use efficiency decreased for both wheat and maize with increasing manure rates. Manure N use efficiency in both the wheat and maize was about 50%, when the cattle manure application rate was 45000 kg/ha (equivalent to N input of 337.5 kg/ha for each crop) (Table 6.2). Ju et al. (2014) reported that average N fertilizer use efficiency for cropping was 50% to 55% globally. The N uptake by crops in the control treatment mainly originated from the soil N pool and N use efficiency in the 15000 kg/ha treatment for the wheat and maize were both over 100%.

6.4. DISCUSSION

The distribution and accumulation of NO_3^--N in the soil profile were influenced greatly by temperature, rainfall, plant uptake, fertilizer application rate and soil type (Ye et al. 2008; Sun et al. 2012; Ji et al. 2014). The experimental results indicate that in the wheat-maize rotation cropping system the NO_3^--N contents and accumulation in the 0–100 cm soil profile increased with increasing cattle manure application rate. This is similar to the results of Yang et al. (2012) and Liu et al. (2015). Yang et al. (2012) found that NO_3^--N accumulation in the 0–120 cm soil profile increased with increasing urea dosage. The current research results showed that differences in NO_3^--N contents in the deep layers of soil (60-100 cm) between manure application rates were significant, with much higher NO_3^--N contents in the treatments with the two highest manure application rates. The accumulation of NO_3^--N in the deep soil layers (out of the root zones) posed a great risk for leaching to groundwater. Some of the accumulated NO_3^--N would have been prone to leaching out of the soil profile when the rainfall season commenced, which might have led to risk of NO_3^- migration and groundwater contamination.

The wheat yield showed a parabolic relationship with the increase in cattle manure application rate, while maize silage yield showed a positive linear correlation with cattle manure application rate. N use efficiency during the wheat and maize seasons was about 50%, when the cattle manure application rate was 45000 kg/ha (equivalent to N inputs of 337.5 kg/ha in each of the wheat and maize seasons). Ju et al. (2014) reported that the average N fertilizer use efficiency, based on the balance method, was 50% to 55% globally. Therefore, manure N use efficiencies in the wheat and maize seasons were similar to the global average level when the cattle manure application rate was 45000 kg/ha (Ju et al. 2014). At the lowest manure application rate, the N uptake by crops exceeded the manure inputs and would inevitably lead to soil N loss and deficiency in the long term. On the other hand, the manure N use efficiency decreased when the manure N rates were over 45000 kg/ha, potentially resulting in N loss through leaching to groundwater. Although animal manure used as an alternative to chemical fertilizer could reduce chemical fertilizer input, excessive application rates may also represent a serious threat to the soil and water environments. Therefore, it is important for agricultural sustainable development that an appropriate application rate for animal manure is used.

CONCLUSION

Wheat and maize silage yields were highest when the cattle manure application rate was 45000 kg/ha/year, with the N use efficiencies of wheat and maize silage both being about 50%. We conclude that a cattle manure application rate of 45000 kg/ha/year (equivalent to 675 kg N/ha/year) would be optimal for this wheat-maize rotation cropping system at the study site in the North China Plain. Furthermore, at this application rate the N use efficiency was appropriate to protect the environment, as above this rate a large amount of N accumulated in the lower soil depths, which created a leaching risk for pollution of groundwater.

ACKNOWLEDGMENTS

This study was funded by the Special Fund for Agro-scientific Research in the Public Interest (201203050-5), Taishan Scholars Program for Overseas Experts, National Key Technology R&D Program of the Ministry of Science and Technology (2012BAD15B02), and Shandong Academy of Agricultural Sciences Youth Fund (2014QNM49).

REFERENCES

Bao S. D. (2000) *Soil analysis*. Agriculture Press of China, Beijing.

Bolan N. S., Adriano D. C., Natesan R. and Koo B. J. (2003) Effects of organic amendments on the reduction and phytoavailability of chromate in mineral soil. *Journal of Environmental Quality* 32: 120-128.

Cordovil C. M., Cabral F. and Coutinho J. (2007) Potential mineralization of nitrogen from organic wastes to ryegrass and wheat crops. *Bioresource Technology* 98: 3265-3268.

Ji Y. Z., Feng W. Z., Hao X. R., Peng Y. J., Han P. H., Ma Z. Y. and Zhang L. J. (2014) Effects of different fertilization pattern on the yield of the rotation system of wheat and maize and soil nitrate accumulation in North China plain. *Ecology and Environmental Sciences (In Chinese)* 23: 1725-1731.

Ju X. T. (2014) The concept and meanings of nitrogen fertilizer availability ratio - Discussing misunderstanding of traditional nitrogen use efficiency. *Acta Pedologica Sinca (In Chinese)* 51: 921-933.

Lin B. (2003) *Chemical fertilizer and non-pollution agriculture*. Agriculture Press of China, Beijing.

Liu M., Song F. P. and Lu Y. Y. (2015) Effects of sulfur-and polymer-coated controlled release urea fertilizers on spatial-temporal variations of soil NO_3-N and nitrogen balance and nitrogen use efficiency. *Journal of Plant Nutrition and Fertilizer (In Chinese)* 21: 541-548.

Liu Q., Zhang B., Xie Y. P., Li J. T., Zhang J. F., Yuan M. M. and Wang D. J. (2008) Leaching characteristics of nitrogen and phosphorus in treated paddy fields with chicken manure. *Chinese Journal of Eco-Agriculture (In Chinese)* 16: 91-95.

Lu R. K. (1999) *Soil agricultural chemical analysis method*. Agricultural science and technology press of China, Beijing.

Salvador P. H., Manuel P. V. and Andres S. A. (2009) hydro-economic modeling framework for optimal management of groundwater nitrate pollution from agriculture. *Journal of Hydrology* 373: 193-203.

Sobańtka A. P., Thaler S., Zessner M. and Rechberger H. (2014) Extended statistical entropy analysis for the evaluation of nitrogen budgets in Austria. *International Journal of Environ Science and Technology* 11: 1947-1958.

Sun M., Meng G. P., Zhang X. L., Feng S. Y. and Huo Z. L. (2012) Characteristics of nitrate-nitrogen distribution and accumulation in soil profiles at intensive planting areas. *Acta Scientiae Circumstantiate (In Chinese)* 32: 902-90.

Wang C. Y., Zhou J. B., Zhen X. F., Li S. X. (2007) Effects of different cultivation methods on soil residual nitrate under winter wheat- summer maize cropping system. *Plant Nutrition and Fertilizer Science (In Chinese)* 13: 991-997.

Yang R. J., Chai S. X. and Ma Z. M. (2012) Dynamic changes of soil moisture and nitrate nitrogen in wheat and maize intercropping field under different nitrogen supply. *Acta Ecologica Sinica (In Chinese)* 32: 7905-7912.

Yao L. X. and Zhou X. C. (2005) Impact of organic manure on the environment and its corresponding preventive researches. *Chinese Journal of Eco-Agriculture* (In Chinese) 13:113-115.

Ye Y. L., Li L. and Sun J. H. (2008) Effect of intercropping three legume crops with maize on soil nitrate-N accumulation and distribution in the soil profile. *Chinese Journal of Eco-Agriculture* (In Chinese) 16: 818-823.

Yuan X. M., Tong Y. A., Yang X. Y., Li X. L. and Zhang F. S. (2000) Effect of organic manure on soil nitrate nitrogen accumulation. *Soil and Environmental Sciences* (In Chinese) 9: 197- 200.

Yu H., Ding W., Luo J., Geng R., Ghani A. and Cai Z. (2012) Effects of long-term compost and fertilizer application on stability of aggregate-associated organic carbon in an intensively cultivated sandy loam soil. *Biology and Fertility of Soils* 48: 325-336.

Zhang X. M., Dong Y. H., Wang H. and Shen D. (2007) Structure of livestock and variation of fecal nitrogen pollution load in China. *Environmental Science* (In Chinese) 28: 1311-1318.

In: Greenhouse Gas Emissions …
Editors: Jiafa Luo and Yan Li
ISBN: 978-1-53611-100-2

Chapter 7

A Review of the Life Cycle Assessment of Off-Grazing Technologies

Stewart Ledgard*[*] and Shelley Falconer
AgResearch, Hamilton, New Zealand

7.1. Introduction

Life Cycle Assessment (LCA) is a systems approach to assess the environmental impacts during the life cycle of products and services (ISO 2006). LCA can be particularly useful in assessing pollution swapping and trade-offs along the product's life cycle or between different environmental impacts (including climate change from greenhouse gas (GHG) emissions). Another benefit of applying LCA is that insight can be gained into the overall environmental impact when farm practices change or mitigation options are implemented.

An LCA study can be conducted across various stages in the production chain of a product or products. For example, pasture-to-plate studies investigate the whole production chain, from the production of fertilisers to the preparation of food by the consumer. Alternatively, a cradle-to-farm gate study can be performed to investigate the total GHG emissions (or carbon footprint)

[*] Corresponding author: Dr Stewart Ledgard, AgResearch, 10 Bisley Road, Private Bag 3123, Hamilton 3214, New Zealand. E-mail: Stewart.Ledgard@agresearch.co.nz.

of products up to the farm gate as a specific point in the production chain. Most published research using LCA methodology for estimating the carbon footprint of dairy products have been limited to the cradle-to-farm-gate stages, partly because these stages are the dominant contributor to the carbon footprint over the whole life cycle of dairy products (e.g., Thoma et al. 2013).

LCA can be used to assess the carbon footprint of off-grazing systems while accounting for the changes that occur as a result of the implementation of these off-grazing technologies (Figure 7.1). Examples of these changes are the use of different feeds, which vary in their carbon footprint, and a reduced use of artificial fertiliser due to greater control over the storage and release of animal effluent onto land. Unfortunately, there have been limited studies that have applied LCA to evaluate off-grazing technologies in pastoral farming systems dominated by grazing of pastures. This is most relevant in New Zealand and other temperate countries such as Australia, Chile, Ireland, Uruguay and South Africa. In contrast, numerous studies have applied LCA in intensive dairy systems dominated by animal housing. In this review, we will focus first on the LCA studies of off-grazing technologies used in pastoral farming systems Secondly, we will review published LCA studies of intensive dairy systems to gain an insight into the hotspots of the carbon footprint of animal housing systems.

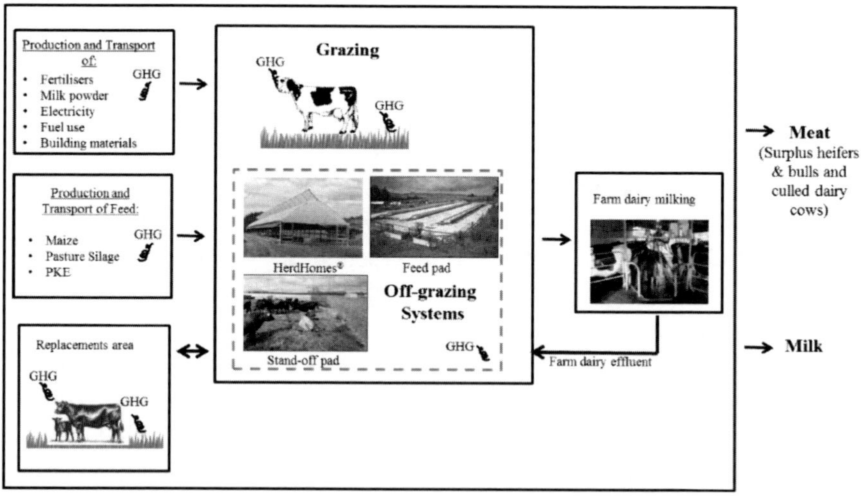

Figure 7.1. A flowchart of the carbon footprint of New Zealand milk production, including multiple off-grazing wintering strategies, covering all stages through to where milk leaves the farm-gate (GHG: greenhouse gases).

There have been a range of published dairy system studies (cradle-to-farm-gate) on the carbon footprint of milk from all-year grazing (Basset-Mens et al. 2009; Gollnow et al. 2014; Chobtang et al. 2015), mixed grazing/housing (Rotz et al. 2010; Bartl et al. 2011; Flysjo et al. 2011; Hagemann et al. 2011; Kristensen et al. 2011; Belflower et al. 2012; O'Brien et al. 2012a; Nguyen et al. 2013; Guerci et al. 2013; Del Prado et al. 2013; de Leis et al. 2015; Wolf et al. 2017) or all-year housing (Wang et al. 2015; Battini et al. 2016) systems. However, few studies have included direct comparison of different production systems and it is not possible to directly compare results from these different studies due to the differences in methodology used. Thus, the next sections are limited to published studies where direct comparisons have been made using the same methodology.

7.2. LCA Comparison of All-Year Grazing and Off-Grazing Systems

Four studies (Flysjo et al. 2011; Hagemann et al. 2011; Belflower et al. 2012; de Leis et al. 2015) included a direct comparison between a grazed dairy farm system and an off-grazing system (either fully housed or partly housed whereby the animals were grazed for a short period over the summer months – the most common practice in Europe). Table 7.1 shows some of the characteristics of the farms in each study and the GHG contribution breakdown.

Flysjö et al. (2011) compared dairy systems in New Zealand and Sweden. In the New Zealand average dairy farm system for the year 2004/05, all-year grazing was assumed and there was a small amount of brought-in-feed (based on DairyNZ survey data). In Sweden, the average dairy farm system for the year 2005 was based on an outdoor period of 2.5 months for milking cows and 5.5 months for heifers, resulting in pasture intake from grazing of less than 10% of the total feed intake. The total GHG emissions for milk at the farm gate were estimated at 1.00 kg CO_2-eq/kg ECM in New Zealand and 1.16 kg CO_2-eq/kg ECM in Sweden. In both cases the largest contributor was animal enteric methane emissions (Table 7.2). Although the methane/kg milk was lower for Sweden, the difference was more than countered by the higher emissions from feed production and energy use.

Table 7.1. Summary of the carbon footprint of energy-corrected milk (ECM) and main contributing gases from research studies comparing all-year grazing, part-grazing and all-year housing (confined) dairy farm systems

Study	System	Country	kg CO_2-eq/kg ECM	Milk Yield (kg ECM/cow)	Contribution (%)		
					CH_4	N_2O	CO_2
Flysjö et al. 2011	Grazing	New Zealand	1.00	4118	63	26	11
	Part-Grazed	Sweden	1.16	8274	50	32	18
Hagemann et al. 2011	Grazing	New Zealand	1.18	4361	57	31	12
	Grazing	Brazil	1.36	3921	65	23	12
	Confined feedlot	USA	0.82	9870	54	22	24
	Forage-based	Germany	1.11	8126	54	25	21
Belflower et al. 2012	Grazing	USA	0.88	5000	nd	nd	nd
	Confined	USA	0.87	10700	nd	nd	nd
de Leis et al. 2015	Grazing	Brazil	0.74	5305	57	16	27
	Confined feedlot	Brazil	0.54	7667	55	22	23
	Part-Grazed feedlot	Brazil	0.78	7324	51	25	24

nd = not determined.

Hagemann et al. (2011) compared the GHG emissions from bovine milk production systems in 38 countries, with a more detailed analysis conducted on 6 farms, which included grazing systems in New Zealand and Brazil, an all-year feedlot system in California and a forage based system in Germany. The total GHG emissions for milk at the farm gate (kg CO_2-eq/kg ECM) were estimated at 1.18 in New Zealand, 1.36 in Brazil, 0.82 in California and 1.11 in Germany (Table 7.1). The underlying farm data for the study were the typical farms of the International Farm Comparison Network (IFCN), which is a worldwide association of agricultural scientists, advisors and dairy farmers. For all four farming systems, enteric methane was the largest contributor to the GHG emissions (Table 7.3). The lowest carbon footprint for USA milk coincided with high milk production per cow and relatively low emissions from feed production.

Table 7.2. Main contributing sources to total GHG emissions from average New Zealand and Swedish dairy farm systems from Flysjö et al. (2011)

Hotspots identified:	Grazing	Part-Grazing
	New Zealand	Sweden
Enteric fermentation	62%	46%
Direct N_2O from excreta/manure	16%	11%
Direct N_2O from N fertiliser application	6%	7%
CO_2 from N fertiliser production	5%	3%
CO_2 from dicsel used	1%	8%

Belflower et al. (2012) simulated two actual dairy farming systems in Georgia in South-eastern USA using the Integrated Farm System Model (IFSM) tool. An intensive pasture-based grazing system was compared to a crop-based confined (free-stall) housing system.

Table 7.3. Main contributing sources to total GHG emissions from dairy farms in different countries based on different systems ranging from all-year grazing to all-year confined housing (Hagemann et al. 2011)

	Grazing	Grazing	Confined feedlot	Forage-based
	New Zealand	Brazil	USA	Germany
Enteric fermentation	55%	61%	52%	51%
Purchased feed	26%	25%	19%	19%
Energy consumption	13%	0%	4%	12%
Fertiliser	2%	6%	2%	9%
Manure	4%	8%	23%	9%

Table 7.4. Predicted annual ammonia emissions for the two dairy farms in Georgia, USA (Belflower et al. 2012)

	Pasture-based dairy	Confinement dairy
Total ammonia volatilisation (kg/cow)	26.4	88.5
In housing facility (kg/cow)	3.3*	40.6
During manure storage (kg/cow)	0	29.2
Following field application (kg/cow)	4.1	10.8
During grazing (kg/cow)	18.9	7.9**
Ammonia N volatilisation (g/kg ECM)	4.3	6.5

*assume this was in the milking parlour.

**dry cows and heifers were placed on pasture to graze.

The carbon footprint of milk was similar for both systems at 0.88 and 0.87 kg CO_2-eq/kg ECM for the pasture-based and confined systems, respectively. The lower methane emissions per kg milk associated with very high milk production per cow from the housing system were balanced out by the much higher N_2O and CO_2 emissions from feed and manure management. One feature of the housing system was the much higher ammonia emissions (an indirect source of N_2O emissions) from the free-stall barn floors and manure storage facilities relative to that from excreta deposition on soil from grazing animals (Table 7.4).

de Leis et al. (2015) compared three different dairy production systems in the southern regions of Brazil; a confined feedlot system, a semi-confined feedlot system (including some grazing) and a pasture-based grazing system. The LCA was performed based on primary data gathered from three different farms representing the average of the years 2008/2009 and 2009/2010. The comparative LCA showed that the confined feedlot system had a lower carbon footprint than the other systems studied (Table 7.5). Total GHG emissions were 0.535 kg CO_2-eq/kg ECM for the confined system, 0.778 kg CO_2-eq/kg ECM for the semi-confined system and 0.738 kg CO_2-eq/kg ECM for the pasture-based system. The contribution of methane from manure management in the part-grazed system was greater than that in the other systems because the manure in the milking stage was handled as liquid slurry in tanks with a longer storage period than the other systems, leading to higher methane emissions due to methanogenesis. Feed production emissions were high in the pasture-based grazing system because they used corn and soybean meal as opposed to cottonseed and brewers grain for the feedlot systems. For the pasture-based system the corn silage and pasture used swine manure as an organic fertiliser which had high GHG emissions associated with it. Thus, the results were highly dependent on the management choices made in each of the systems.

Table 7.5. Main contributing sources to total GHG emissions from different dairy farm systems in Brazil (de Leis et al. 2015)

	Pasture-based grazing	Part-grazing + Feedlot	Confined Feedlot
	Brazil	Brazil	Brazil
Enteric fermentation	50%	47.6%	52.3%
Feed production	45.9%	36.9%	37.8%
Manure management	3.2%	13.7%	7.8%

Direct comparison of the carbon footprint of milk across the four studies cannot be made due to methodology differences. However, they can be used to examine effects of grazing versus partial or full housing systems. The all-year grazing systems ranged from being higher to being lower than that from the partial or full housing systems. In the three studies with full housing systems, two of the studies had a lower carbon footprint for milk than the grazing system. This was associated with much higher milk production per cow, resulting in relatively lower enteric methane emissions per kg milk. However, the results were highly dependent on the corresponding feed and manure management systems, and in the USA study of Belflower et al. (2012) these differences balanced out resulting in the same milk carbon footprint in the housing and grazing systems.

7.3. LCA of All-Year Housing and Part-Grazing Systems

Five northern hemisphere studies compared an all-year housing dairy farm system with a housing system that included grazing of pasture for part of the year (Arsenault et al. 2009; Rotz et al. 2010; O'Brien et al. 2012; Guerci et al. 2013; Wolf et al. 2017).

Table 7.6. Contribution analysis to the total carbon footprint of milk for a grass-based and a confinement Irish dairy farm (O'Brien et al. 2012b)

	Grass-based dairy farm	Confinement Dairy farm
Milk/cow (kg FPCM[1])	6639	8040
Total carbon footprint (kg CO_2e/kg FPCM)	0.874	1.027
Of which:		
Enteric fermentation (CH_4)	45.6%	35.6%
Production of fertiliser applied to pasture (CO_2+N_2O)	11.6%	0.1%
Fertiliser application to pasture (N_2O)	8.2%	N/A
Manure deposited by grazing cattle on pasture (N_2O)	7.7%	0.7%
Manure storage and cattle housing (CH_4+N_2O)	4.0%	11.9%
Electricity for animal operations e.g., milking (CO_2)	1.9%	5.9%
Production of concentrates (mainly CO_2+ N_2O)	4.8%	20.5%
Land use change during production of concentrates (mainly CO_2)	1.2%	11.3%

[1] Fat and Protein Corrected Milk.

In Ireland, O'Brien et al. (2012a, b) compared the environmental impacts of an all-year confinement housing dairy farm with a seasonal pasture-based dairy farm. Within this confinement system, cows were housed year-round but the replacement animals were only indoors during the winter period. The carbon footprint was greater for the confinement dairy system (1.027 kg CO_2-eq/kg Fat and Protein Corrected Milk (FPCM) vs. 0.874 kg CO_2-eq/kg FPCM for the grass-based system), irrespective of whether the GHG emissions were expressed per unit of milk or per unit area. The greater carbon footprint was due to higher emissions associated with the greater use of concentrates as feed and the longer manure storage period (Table 7.6). Average milk production per cow was higher (+21%) in the confinement system than in the pasture-based system.

In the USA, Rotz et al. (2010) applied an LCA approach to study the carbon footprint of milk from four dairy systems in Pennsylvania and two dairy systems in California (Table 7.7). The four simulated representative farms in Pennsylvania were: a) '60 cow confined' which referred to 60 cows plus replacement heifers housed in a tie-stall barn and fed a high-forage diet, b) '60 cow grazing', for 60 cows plus replacement heifers housed in a tie-stall barn while grazing rotationally on perennial grass pasture during 6 months of the year; with 65% of their forage from pasture, c) '500 cow confined' which referred to 500 cows plus replacement heifers housed in a free-stall barn and fed a high-grain diet, and d) '2,000 cow drylot' for 2,000 cows in a free-stall barn with a drylot system and fed a high-grain diet. The two simulated representative farms in California were: e) '500 cow confined', for 500 cows plus replacement heifers housed in a free-stall barn, and f) '2,000 cow drylot' for 2,000 cows in a free-stall barn with drylot system.

The carbon footprint of the 60 cow confinement and part-grazing systems were similar at 0.69 and 0.62 kg CO_2e/kg ECM (Table 7.7). Higher housing methane emissions from the confinement system (including from bedding) were countered by higher crop/forage N_2O emissions from animal urine during gazing of grass in the part-grazing system. Emissions from fuel combustion were 30% lower in the part-grazing system due to fewer machinery operations and lower secondary emissions with reduced use of equipment, fuel, and electricity.

The carbon footprint of the 500 cow confined system was lower at 0.53 kg CO_2e/kg ECM. The use of a free-stall barn and more grain in the diet resulted in 38% lower methane emissions per cow from the animal and housing sources compared with the 60-cow confinement forage system (a). However, the use of long-term manure storage greatly increased the GHG emissions during manure handling. Emissions from fuel combustion and secondary sources also increased slightly with these feeding and manure handling strategies. Overall, the net GHG emission per cow was 10% less than that from system (a) and with the greater milk production per cow, the carbon footprint was reduced by 23%.

The carbon footprint of the 2,000 cow drylot system was lowest at 0.46 kg CO_2e/kg ECM. Compared with the 500 cow confined system, use of the drylot system reduced methane and N_2O emissions from the manure storage. Overall, GHG emission per cow and the carbon footprint of milk were 13% less than that of the large confinement system (c) with the same milk production and similar feeding strategy.

Table 7.7. Contribution analysis to the total carbon footprint of milk from dairy systems in Pennsylvania, USA (Rotz et al. 2010)

	60 cow Confined (forage diet)	60 cow Grazing (forage+grass)	500 cow Confined (grain diet)	2,000 cow Drylot (grain diet)
Milk/cow (kg ECM[1])	8500	8500	10000	10000
Total carbon footprint[2] (kg CO_2e/kg ECM) Of which[3]:	0.69	0.62	0.53	0.46
Animals and housing (CH_4)	37.7%	34.9%	23.1%	24.8%
Animals and housing (CO_2)	40.0%	40.6%	42%	44.7%
Cropland (N_2O)	9.6%	14.4%	7.6%	8.0%
Secondary sources[4]	8.3%	7.6%	9.2%	7.7%
Manure storage (CH_4+N_2O)	0%[5]	0%	14.0%	5.7%

[1] Energy Corrected Milk.

[2] In this total carbon footprint a C net balance is included which accounted for the C fixed during crop growth and carbon dioxide from animals and manure.

[3] Contributions are given relative to the total GHG emissions (in CO_2-eq per cow), before the positive impact from croplands (e.g., through CO_2-assimilation) is subtracted to get the total carbon footprint estimate in CO_2-eq per kg ECM.

[4] Defined as GHG emissions associated with the production of fuel, electricity, machinery, fertiliser, pesticide, plastic, and maintenance of animals and handling of manure.

[5] No storage emissions since manure was spread daily.

Table 7.8. Comparison of three dairy farm systems in Germany that varied in level of pasture grazing for the carbon footprint of milk, production and sources of feed used (Wolfe et al. 2017)

	Zero-grazing	Restricted-grazing	Unrestricted-grazing
Milk yield (kg ECM/cow/year)	10036	10068	10042
Total carbon footprint (kg CO_2e/kg ECM)	1.07	1.05	1.06
Grazing days (April – October)	0	182	182
Hours per day grazing	0	10	20.5
Feed intake (kg DM/year)	7292	7228	7236
Grassland based feed, except pasture	27%	15%	14%
Pasture	0	22%	38%
Field grass silage	11%	8%	6%
Maize silage	26%	21%	13%
Sugar beet pulp	7%	4%	4%
Soybean meal	1%	1%	1%
Rapeseed meal	6%	5%	3%
Triticale	5%	11%	14%
Dairy concentrate	16%	13%	8%

Rotz et al. (2010) concluded that in terms of the carbon footprint of milk, the 2,000 cow drylot system was the most efficient, followed by the 500 cow confinement system and that the carbon footprint in these systems could be further reduced by improving manure storage. However, the study did not include consideration of other environmental impact indicators and the high ammonia emissions from the drylot system may have a negative effect on those indicators influenced by ammonia.

In a recent study of German dairy farm systems, Wolf et al. (2017) used a representative dairy farm system using German average data for crop yields and fertiliser rates (Haenel et al. 2014) to model the effects of three grazing systems (zero, restricted and unrestricted grazing). The dairy cows and heifers were loosely housed in stables with cubicles, slatted floors, and slurry-based manure management, being the most common system in Germany. It was assumed that the slurry from all cattle was untreated and stored in a slurry tank with a natural crust. The carbon footprint of milk for all three systems was similar at 1.07, 1.05 and 1.06 kg CO_2e/kg ECM for the zero, restricted and unrestricted grazing systems, respectively (Table 7.8). For the zero-grazing system, methane contributed 59% to the total emissions, N_2O 28% and CO_2

13%. The other two grazing systems followed a similar pattern for the relative contributions. The difference between the results was that the high emissions from crop cultivation (higher in the zero-grazing system) were interchanged for the high direct N_2O emissions from excreta during grazing in the restricted and unrestricted grazing systems, respectively. There were some differences in feed types used between the systems (Table 7.8) and this will have had some effect on overall carbon footprint results.

The environmental impacts from a range of different case-study representative dairy farming systems across Europe were studied by Guerci et al. (2013), which included comparison of an all-year housing system and a system with some pasture grazing in Germany (Table 7.9). All five Italian farms were all-year housing systems and illustrate the variability in carbon footprint of milk (almost two-fold range) that can occur due to management and site factors within a similar type of farming system. The Danish farms were also similar, although with a small variation in the component of grazing, and showed variability in the carbon footprint of milk. For the two German farm systems, the carbon footprint of milk for the part-grazing system (where 71% of the diet was pasture) was less than half that for the all-housing system, despite the lower milk production per cow. The higher carbon footprint of milk from the all-year housing system was largely due to the relatively high emissions associated with production and use of feeds and bedding. This was also associated with higher values for Eutrophication (+67%) and Acidification (+143%) indicators for the all-year housing system compared to the part-grazing system.

Arsenault et al. (2009) applied a comprehensive LCA approach to assess multiple environmental impacts of a Canadian (Nova Scotia) year-round confinement-based dairy system and a pasture-based dairy system. The latter involved the cows spending 150 days on pasture and 215 days in confinement. The milk production of cows in the pasture-based system was 20% less than that of cows in full-confinement, due to higher feed supplementation in the confinement system. The systems were compared for their impact on: Abiotic depletion, Global warming potential, Ozone layer depletion, Human toxicity, Freshwater aquatic ecotoxicity, Terrestrial ecotoxicity, Photochemical oxidation, Acidification, Eutrophication, Land use, and Cumulative energy demand. The carbon footprint of milk (kg CO_2e/kg) for the all-housing and part-grazing systems was 0.974 and 0.988, respectively. Hotspots in the carbon footprint for both systems were methane from enteric fermentation, methane from manure storage, and N_2O from manure storage. Provision (including production and transport) of concentrates contributed 25.5% in the confinement-based system and 21.4% in the pasture-based system. Forage production was also a hotspot in the confinement (24%) and pasture-based system (29%), in particular the production of silage. Arsenault et al. (2009) concluded that grazing cows for five months per year (typical of pasture systems in Nova Scotia) had little effect on overall environmental impact. However, they carried out scenario modelling of the systems and indicated that extending the grazing period to seven-months resulted in lower impacts from the part-grazing system for seven of the eleven environmental impact categories evaluated, with greatest potential improvements associated with acidification potential, ozone depletion potential, human toxicity and fresh water ecotoxicity. There was little change in the carbon footprint of milk with seven months grazing (0.98 kg CO_2e/kg). In contrast, land use was the only category that was much higher for the pasture-based system due to an increased land area requirement for pasture production. It was concluded that in Nova Scotia, Canada, the transition to a full confinement dairy system does not result in environmental benefits, as some of the modelled impacts were higher in the confinement system, while there was little difference in the carbon footprint of milk.

Table 7.9. Carbon footprint of milk and contributing factors for a range of dairy farm systems in Germany (GER), Denmark (DK) and Italy (IT) (Guerci et al. 2013)

	GER-1	GER-2	DK-1	DK-2	DK-3	IT-1	IT-2	IT-3	IT-4	IT-5
Milk per cow (kg ECM/year)	10964	6277	8527	10427	7976	10222	6330	9391	10481	7891
Pasture (%DM intake)	0	71	6	7	0	0	0	0	0	0
Stocking rate (LU/ha)	2.1	1.1	1.2	1.2	2.1	2.2	2.5	5.6	9.8	11.0
Total carbon footprint (kg CO_2-eq/kg ECM). Of which:	1.32	0.55	1.57	1.27	1.66	1.36	1.91	1.47	1.18	1.11
Enteric fermentation	32%	75%	29%	34%	31%	35%	38%	38%	38%	45%
Manure storage	17%	11%	15%	18%	15%	16%	17%	16%	17%	22%
Farm crop	18%		28%	19%	21%	15%	22%	14%	12%	11%
Farm energy	6%	8%	11%	16%	8%	9%	9%	5%	7%	5%
Purchased feed and bedding	17%	6%	6%	5%	21%	20%	7%	23%	23%	17%
Fertilisers &pesticides	10%	-	11%	8%	4%	5%	7%	4%	3%	-

CONCLUSION

A limited number of LCA studies have included a comparison of dairy farm systems that covered some of the variation between all-year grazing and all-year housing systems. These showed no clear differences between the different types of systems and the variation between farms with a single system (e.g., all-year housing; Guerci et al. 2013) was at least as large as any apparent system-related differences.

Most all-year housing systems included significant use of cereals or concentrates and had higher milk production per cow. This was associated with lower enteric fermentation per kg milk, which was the main single source of GHG emissions in both housing and grazing systems. Other hotspots in the housing systems were methane and N_2O emissions from manure storage or management (Chapters 2-3) and GHG emissions associated with the production of concentrates. In contrast, hotspots in the pasture-based systems were N_2O emissions from animal excreta on pasture during grazing, GHG emissions associated with the production of fertiliser and N_2O emissions following N fertiliser application.

An important determinant of the carbon footprint of milk from all-year housing systems, was the relative difference between the reduced enteric methane emissions and increased manure and feed GHG emissions within the housing systems. Housing systems with very high milk production per cow and efficient manure and feed management practices generally had a relatively low carbon footprint of milk. For grazing systems, a low carbon footprint could be related to systems with high use of quality pastures and lower fertiliser N inputs.

Few comparative system studies covered more than carbon footprinting. When an LCA study is expanded to cover more environmental impacts than just climate change, the relative effects of housing or grazing systems can vary for the different environmental impacts. For example, Arsenault et al. (2010) estimated higher impact results for a number of other indicators for an all-year housing system than a housing system with a seven-month grazing component. The relative impacts will depend on various factors including the method of housing and of restricted or off-grazing used, as well as the level of milk production and manure management system.

More research is needed to get a complete picture of the total environmental impact of typical off-grazing technologies compared to that of all-year grazing systems, such as that implemented in New Zealand. This requires the use of a common LCA methodology across the different farm systems.

Acknowledgments

The authors acknowledge funding from the Ministry for Primary Industries of New Zealand (MPI) through the Sustainable Land Management and Climate Change (SLMACC) programme.

References

Arsenault N., Tyedmers P. and Fredeen A. (2009) Comparing the environmental impacts of pasture-based and confinement-based dairy systems in Nova Scotia (Canada) using life cycle assessment. *International Journal of Agricultural Sustainability* 7: 19-41.

Belflower J.B., Bernard J.K., Gattie D. K., Hancock D.W., Risse L.M. and Rotz C. (2012) A case study of the potential environmental impacts of different dairy production systems in Georgia. *Agricultural Systems* 108: 84-93.

de Léis C.M., Cherubini E., Ruviaro C.F., Prudêncio da Silva V., do Nascimento Lampert V., Spies A. and Soares S.R. (2015) Carbon footprint of milk production in Brazil: a comparative case study. *International Journal of Life Cycle Assessment* 20: 46-60.

Flysjo A., Henriksson M., Cederberg C., Ledgard S. and Englund J.E. (2011) The impact of various parameters on the carbon footprint of milk production in New Zealand and Sweden. *Agricultural Systems* 104: 459-469.

Guerci M., Knudsen M.T., Bava L., Zucali M., Schonbach P. and Kristensen T. (2013) Parameters affecting the environmental impact of a range of dairy farming systems in Denmark, Germany and Italy. *Journal of Cleaner Production* 54: 133-141.

Haenel H.D., Rösemann C., Dämmgen U., Freibauer A., Döring U., Wulf S., Eurich-Menden B., Döhler H., Schreiner C. and Osterburg B. (2016) Calculations of gaseous and particulate emissions from German agriculture 1990-2014: *Report on methods and data (RMD) submission*

2016 [Berechnung von gas-und partikelförmigen Emissionen aus der deutschen Landwirtschaft 1990-2014: Report zu Methoden und Daten (RMD) Berichterstattung 2016] (No. 39). Johann Heinrich von Thünen Institute, Federal Research Institute for Rural Areas, Forestry and Fisheries.

Hagemann, M., Hemme, T., Ndambi, A., Alqaisi, O. and Sultana, M. N. (2011) Benchmarking of greenhouse gas emissions of bovine milk production systems for 38 countries. *Animal Feed Science and Technology* 166-167: 46-58.

ISO 14040 (2006) *Environmental management-life cycle assessment-principles and framework.* International Organization for Standardization, Geneva.

O'Brien D., Shalloo L., Patton J., Buckley F., Grainger C. and Wallace M. (2012a) Evaluation of the effect of accounting method, IPCC v. LCA, on grass-based and confinement dairy systems' greenhouse gas emissions. *Animal* 6: 1512-1527.

O'Brien D., Shalloo L., Patton J., Buckley F., Grainger C. and Wallace M. (2012b) A life cycle assessment of seasonal grass-based and confinement dairy farms. *Agricultural Systems* 107: 33-46.

Rotz C.A., Montes F. and Chianese D. S. (2010) The carbon footprint of dairy production systems through partial life cycle assessment. *Journal of Dairy Science* 93: 1266-1282.

Thoma G., Popp J., Nutte, D., Shonnard D., Ulrich R., Matlock M., Kim D.S., Neiderman Z., Kemper N., East C. and Adom F. (2013) Greenhouse gas emissions from milk production and consumption in the United States: A cradle-to-grave life cycle assessment circa 2008. *International Dairy Journal* 31: S3-S14.

Wolf P., Groen E.A., Berg W., Prochnow A., Bokkers E.A.M., Heijungs R. and de Boer I.J.M. (2017) Assessing greenhouse gas emissions of milk production: which parameters are essential? *International Journal of Life Cycle Assessment* 22: 441-455.

In: Greenhouse Gas Emissions …
Editors: Jiafa Luo and Yan Li
ISBN: 978-1-53611-100-2

Chapter 8

THE ESTIMATED GREENHOUSE GAS EMISSIONS FROM DIFFERENT TYPES OF HOUSED/RESTRICTED-GRAZING SYSTEMS USED DURING THE WINTER IN SOUTHLAND AND WAIKATO, NEW ZEALAND, USING LIFE CYCLE ASSESSMENT

Stewart Ledgard*[*] *and Shelley Falconer
AgResearch, Hamilton, New Zealand

8.1. INTRODUCTION

Chapter 7 reviewed published dairy farm system research on the effects of different management practices (from grazing through to housing) on total GHG emissions (carbon footprint of milk) using life cycle assessment (LCA). This chapter presents detailed case studies on the carbon footprint of milk, comparing the effects of different farm system scenarios in winter involving grazing, stand-off pads or housing systems in two contrasting regions of New Zealand (Waikato and Southland).

[*] Corresponding author: Dr Stewart Ledgard, AgResearch, 10 Bisley Road, Private Bag 3123, Hamilton 3214, New Zealand. E-mail: Stewart.Ledgard@ agresearch.co.nz.

Dairy farming in New Zealand is predominantly based on grazing of perennial grass/clover pastures, e.g., an average of 80-85% of feed intake by dairy cows is from grazed pasture (Clark et al. 2013). Most New Zealand farms have a seasonal milk supply system, with cows calving in early-spring so that the period of greatest pasture growth coincides with the highest feed demand by lactating cows. The cows are non-lactating during winter. The Waikato region is in the North Island of New Zealand and has a temperate/ subtropical climate with moderate winter pasture growth (c. 15-25 kg dry matter (DM)/ha/day, DairyNZ 2016), whereas Southland is in the southern South Island and has low winter growth of pastures (c. 5-15 kg DM/ha/day, DairyNZ 2016). Low winter pasture growth in Southland means a greater requirement for use of conserved or purchased feeds and is a better fit with the use of winter housing/stand-off systems.

8.2. Methods

8.2.1. Functional Unit and System Boundary

The functional unit of the study was one kg of fat- and protein-corrected milk (FPCM) calculated using the IDF (2015) equation. The total GHG emissions were allocated between the co-products milk and live-weight sold for meat using the IDF (2015) methodology.

The system boundary in this study was the “cradle-to-farm-gate”. It covered the main processes for the on-farm production and extraction of milk, as well as all related off-farm processes (Figure 7.1 in Chapter 7). On-farm processes included the production and utilisation of pasture (thus determining methane emissions from animals and nitrous oxide emissions from their excreta), the use of farm machinery (representing diesel and petrol use) and the use of equipment for milk extraction, farm dairy effluent management and water supply (determining electricity use). The systems for managing the animal excreta and manure from the stand-off pads and animal housing were also included. The model included the off-farm production of pasture for the dairy cow replacements and the production and delivery of inputs (e.g., fertilisers) to the crops and pasture. The additional capital required for the restricted grazing systems was also modelled in the LCA, including emissions from production and maintenance of infrastructure (e.g., animal housing, stand-off pads). The above components have been found, in previous studies (e.g., Ledgard et al. 2008), to account for at least 99% of the likely life cycle GHG emissions from the cradle-to-farm-gate.

8.2.2. Description of Systems

Average dairy farm systems in Waikato and Southland (Table 8.1) were based on regional survey farm data for 2010/2011 from LIC (2011). Additional specific data for main farm inputs (e.g., fertilisers, brought-in feeds) was derived from Dairybase (DairyNZ 2015), while lime use was based on modelled requirements (Sinclair et al. 1993). Seasonal milk supply was assumed, with a 70 day non-lactating period over winter. Replacement animals were assumed to be grazed off-farm between weaning and three weeks before calving on pasture, with the system and inputs based on the MAF intensive finishing monitor farm (MAF 2011).

Three management scenarios were modelled for each regional farm, with the all-year grazing scenario and the scenario with wintering cows in barns using a slatted floor (and manure collection in a concrete-under-floor bunker) being in common for both regions (Table 8.2). In Waikato, a winter stand-off system was also modelled, based on eight hours of grazing and sixteen hours on the pad each day. In Southland, the third scenario was a housed system with a solid floor and a deep wood chip mix covering to contain the effluent. Cows were grazed on pasture during all non-winter months of the year for all scenarios.

Table 8.1. The key average annual farm data for the two New Zealand regions studied

	Waikato	Southland
Average area, effective ha	110	206
Cow live weight, kg	445	465
Annual milk yield, litres/cow	3703	4179
Fat and protein corrected milk, kg/ha	12150	12520
Butterfat (% m/v)	5.0	4.9
Protein (% m/v)	3.7	3.8
Dairy cows in milk/ha	2.88	2.63
Replacement rate, %	21.1	21.1
N fertiliser, kg N/ha	126	112
P fertiliser, kg P/ha	44	25
K fertiliser, kg K/ha	40	11
Lime, kg/ha	90	90

Table 8.2. The different cow wintering scenarios, their location and daily hours off pasture during the 10 week wintering period

Waikato	Southland	Daily hours off pasture
Grazing	Grazing	0
Stand-off pad		16
Slatted floor barn	Slatted floor barn	24
	Solid floor barn	24

The monthly feed requirement of animals was calculated using an energy requirement model, as used for the New Zealand GHG Inventory (Clark et al. 2003). No adjustment was made for possible differences in feed requirement between grazed and housed cows. The winter feed requirements were assumed to be partly met by pasture, based on the pasture growth rates for Ohaupo (Waikato) and Winton (Southland), respectively (DairyNZ 2016). The difference was assumed to be met from pasture silage.

In all scenarios, the cows would receive the equivalent of 2447 kg pasture silage DM/ha in Southland and 1771 kg pasture silage DM/ha in Waikato over the 10 week period. It was assumed that cows on the stand-off pad scenario would obtain the same pasture intake during their eight-hour grazing as that of the full-time grazing animals.

For the housed cows, it was assumed that they received the same intake of pasture over the winter period as for the grazing cows, based on a simple cut and carry system to harvest and supply the pasture grown over the ten-week wintering period. Thus, the harvested pasture, supplemented by pasture silage, was calculated to be sufficient to meet the energy requirements of the housed animals.

8.2.3. Methane Emissions during the 70-Day Winter Period of Scenario Differences

The enteric methane (CH_4) emissions were calculated as a product of the amount of feed consumed using the New Zealand-specific emission factor of 21.6 g CH_4 per kg DM (Clark 2001).

The calculation of the methane emissions from faecal deposition to pasture and from the housed systems required different methodologies. In the stand-off pad scenario, it was assumed that 1/3 of the excreta was deposited to pasture with the remainder (2/3) going to the stand-off pad (in proportion to the time spent grazing or on the stand-off pad). The other confinement scenarios had all of the excreta from the 70-day period deposited on barn surfaces and collected to storage. The New Zealand IPCC emission factor of 0.98 g CH_4/kg faecal DM was used for dung from grazing animals deposited on pasture (MfE 2016).

The methane emission from storage of faecal material produced during animal housing was calculated using the IPCC 2006 Tier 2 methodology. The amount of volatile solids were calculated using a simplified version of the IPCC equation 10.24:

VS, kg = DM intake, kg x (1-Digestibility+ UE)*(1-ASH)

where:

- UE is urinary energy fraction (assumed to be 0.04)
- ASH is the ash content of manure calculated as a fraction of the DM feed intake (assumed to be 0.08)
- Digestibility was the average digestibility proportion of the feed

The methane emissions from the stored excreted volatile solids were calculated using the equation:

Methane EF = Bo x 0.67 x MCF

where:

- Bo is the maximum methane producing capacity for manure. The IPCC (2006) report has dairy cattle in Oceania producing 0.24 m^3 CH_4/kg VS. Applying the conversion factor of 0.67 kg CH_4/m^3 gives 0.16 kg CH_4/kg VS.
- MCF is the methane conversion factor (%) for a specific manure management system and its climatic region. An MCF value of 19% was given by IPCC (2006) for the category "pit storage below animal confinements of more than 1 month" at an average temperature of 11°C and for the liquid/slurry category without a natural crust cover. Thus, a value of 19% was adopted based on the above MCF categories.

Putting the above values into the methane EF equation gives 0.0305 kg CH_4/kg VS (i.e., 0.24 m^3 CH_4/kg VS x 0.67 kg CH_4/m^3 CH_4 x 19% = 0.0305 kg CH_4/kg VS).

8.2.4. Gaseous Nitrogen Emissions from Excreta during the 70-Day Winter Period

The nitrogen (N) emissions from excreta during the period off pasture were calculated based on the report of Luo and Longhurst (2008). It was assumed that the amount of N excreted by the cows during this non-lactating period was equivalent to the N intake. The justification for this methodology was that the N contained in milk and meat was accounted for in the calculations of the emissions from N excreted during the grazing/milking period and by the off-farm replacement N emissions.

Emission factors for N gases emitted during manure deposition, as described in Chapter 2, were calculated and are presented in Table 8.3.

During the confinement period the manure was assumed to be all stored as slurry, with the exception of the stand-off pad where it was assumed that 10% of the excreta went to an effluent holding pond and 90% of excreta was held as slurry. The emission factors for the manure storage are shown in Table 8.4.

The calculations allowed for the assumption that the excreta was stored for an additional 2 months beyond the 70 days (removed in October for land application). The emission factors during manure application to soil were NH_3 10% and N_2O 1% of the total N in manure collected in the different confinement systems.

Table 8.3. Gaseous emissions as % of total N excreted on the deposition surface

		Form of N loss from excreta		
Winter system	Surface material	NH_3	N_2O	N_2
Stand-off pads	C-rich material floor	2%	0.01%	0.01%
Animal shelters:				
Solid floor barn	C-rich material	7%	0	0
Slatted floor barn	No lining material	9%	0	0

Table 8.4. Gaseous emissions as % of total N in manure during storage

	NH_3	N_2O	N_2
Holding ponds	15%	0.10%	5%
Slurries	60%/year*	0.1%/year*	2%/year*

* For slurries and solids, the period stored needs to be calculated, i.e., emission rate x months/12.

Table 8.5. Assumed confinement areas per cow and lifespan of the structures

	m^2/cow	lifespan, years
Slatted floor barn	3.3	25
Solid floor barn	4.8	25
Stand-off pad	5	15

8.2.5. GHG Emissions from Infrastructure

The emissions from production and transport of materials used in capital structures (infrastructure) were also calculated. There was no attempt to account for electricity and fuel used on site during the construction apart from energy for relevant excavations. The amounts of materials, per m^2, used in construction were based on case farms (Tony van der Weerden, pers. comm). The total emission from construction was assumed to be spread over the lifespan of the structures (Table 8.5). Thus the capital emission for the year of this study was equivalent to the total capital GHG emissions divided by the lifespan of the structure.

The slatted floor barn and solid floor barn areas were based on observations from existing systems in Southland (Tony van der Weerden, pers. comm.). The area assigned to the stand-off pad was based on 5 m^2/cow for long term (>12 hours/day for more than 3 days in a row). With the emissions from the capital structures at only 0.14-0.36% of the total GHG emissions there was little point in refining the parameters used in the capital calculations further.

The emissions from the removal and relocation of the stored excreta from the slatted floor barn and solid floor barn were based on Southland case study data for each of these systems (Tony van der Weerden, pers. comm.). The energy use for removal of excreta on the stand-off pad was estimated. An emission for the annual excavation and removal of the top 0.1 m of the pad was calculated. The emissions for wood chip production, transport and replacement were also included.

8.2.6. GHG Emissions during the Grazing Period

The GHG emissions for the whole farm system were included in the carbon footprint calculation. Thus the period of grazing for the remaining 295 days was included in the calculations and was the same across all scenarios within each region (Table 8.6). It was assumed that pasture production in the period outside the 70-days would meet the cow's energy needs. The amount of feed required was calculated using the Tier 2 approach of New Zealand IPCC inventory (Clark et al. 2003), which takes into account the energy required to meet the levels of livestock production.

Table 8.6. The key emission factors and parameters according to the IPCC-New Zealand inventory used to calculate GHG emissions during the 295 day grazing period for lactating cows

Process	Mean	References for mean
Kg of CH_4 emitted per tonne of DM ingested	21.6	Clark (2001)
Kg of N_2O-N emitted per kg of N:		
Due to urea-N fertiliser use (EF1)	0.0048	MfE (2016)
Due to urine deposited during grazing (EF3) Due to dung deposited during grazing (EF3)	0.01 0.0025	Kelliher et al. (2005), MfE (2016)
Due to excreta stored in an anaerobic lagoon (EF3 AL)	0.001	IPCC (2006)
Due to atmospheric deposition of NH_3 (EF4)	0.01	IPCC (2006)
Due to leaching and runoff of NO_3^- (EF5)	0.0075	IPCC (2006)
Kg of NO_3-N leached per kg of N excreted or N fertiliser applied (FRACLEACH)	0.07	Thomas et al. (2005)
Kg of NH_3-N emitted per kg of N excreted (FRACGASM)	0.10	MfE (2016)
Kg of NH_3-N emitted per kg of N fertiliser applied (FRACGASF)	0.10	MfE (2016)

Dry matter intakes were multiplied by the average New Zealand pasture N concentration (Ledgard et al. 2003) to calculate the N intake. The N in milk was then subtracted from total N intake by cows in order to calculate the amount of N excreted.

The slatted floor barn scenario used the IPCC (2006) equation for CH_4 emissions from dairy effluent ponds instead of the current New Zealand GHG Inventory equation, i.e.,

$$M = Q * EF$$

where

M = CH_4 emission from manure management (kg CH_4/year)
Q = number of lactating dairy cows
EF = annual CH_4 emission factor for dairy cows, kg CH_4/animal/yr

and

$$EF = VS * T * Bo * 0.67\ kg/m^3 * (MCF/100) * MS$$

where

VS = kg VS/cow/day (Oceania default for 500 kg cow is 3.5 kg VS/cow/day, IPCC 2006)
T = average length of lactation season of 280 days
Bo = 0.24 m^3 CH_4/kg VS added (IPCC default)
0.67 = conversion factor of m^3 CH_4 to kg CH_4
MCF = CH_4 conversion factor for lagoon; IPCC default is 71% (at 13°C average annual temperature) – representing the degree to which Bo is achieved
MS = fraction of total manure mass handled, based on average time fraction per day spent by cows on the milking platform. An MS value of 15% was used based on Laubach et al. (2013).

For all housing scenarios, the following was assumed:

1. An allowance for waste milk entering the anaerobic pond. Estimates of the various milk wastes that could enter the pond were based on Chung et al. (2013). Milk vat contamination was assumed to have a

frequency of once every 3 years. It was assumed that 100% of the milk containing antibiotics would be sent to the ponds. About 22 litres milk per cow/year were assumed to be lost in washing and purging milking machines.

2. An allowance for waste feed materials entering the ponds. The estimation of the methane emission from the silage feed waste assumed a volatile solid content of 94.4% total solid and 0.326 m^3 CH_4/kg VS (Chung et al. 2013). It was assumed that best practices were used and 10% of pasture silage fed in bins was wasted (DairyNZ 2010).
3. Allowance for NH_3 volatilisation at all stages of the manure management. An emission rate of 2 g NH_3-N/cow/hour from deposition of the manure in the cow shed was adopted (Chapter 2). The N_2O gas emission at this site was assumed to be negligible. The emission of NH_3-N from the farm dairy effluent (FDE) in the anaerobic pond was assumed to be 30% of total excreta N in the pond. No direct N_2O emission was assumed to occur from the anaerobic pond.

8.2.7. GHG Emissions Associated with Replacement Animals

It was assumed that an average intensive beef farm was used for grazing replacements, based on the 2011 MAF Intensive Beef Monitor Farm (MAF 2011). The feed requirements for replacement animals were assumed to be met by the on-site production of pasture and pasture silage. Current practices were assumed for the use of inputs for the production of pastures and silages (Basset-Mens et al. 2008). The emissions from the transport of the replacement animals to and from the grazing site were included.

The methodology used for the calculation of GHG emissions from the replacement animals reared off farm was similar to that used for the milking/grazing cows, with a few exceptions. For instance, calculation of the amount of N excreted by the replacements was obtained by subtracting the N contained in increased animal live-weight (associated with the increase in animal growth) from the N intake.

8.2.8. GHG Emissions from Supplementary Feed Production

The pasture silage was assumed to come from the beef farm where replacement animals were grazed. For a sensitivity analysis, the inventory for the production of palm kernel expeller (PKE) was based on Schmidt (2007), emission factors from IPCC (2006) and the Blonk (2014) land use change tool. Economic and country-of-origin data were used for the allocation of the GHG emissions between the palm oil industry co-products.

8.2.9. Calculating Total GHG Emissions

The carbon footprint (equivalent to Global Warming Potential) for a 100 year time horizon (GWP100) was calculated according to IPCC (2007) in kg CO_2-equivalents (CO_2e), i.e., with characterisation factors of CO_2 1, N_2O 298 and CH_4 25. These characterisation factors correspond to the impact of the gases on the heat radiation absorption of the atmosphere. All operations were parameterised in LCA software (SimaPro Version 8.2) and background emissions were based on use of data from ecoinvent (Version 3.2) adjusted for New Zealand-relevant parameters. Fertiliser manufacturing emissions were based on New Zealand specific production (Ledgard et al. 2011).

8.3. Results

8.3.1. Waikato Dairy Farm

The total carbon footprint of milk was lowest for the all-year grazing scenario (Table 8.7). The carbon footprint was 4.9% higher for the stand-off pad scenario (with cows spending 16 hours/day on the pad during winter) and 7.3% higher for the slatted floor barn scenario.

The GHG emissions associated with the 70 day wintering period contributed 16% to the total carbon footprint for the all-grazing scenario, with 72% coming from the remainder of the year (295 days including lactation period) and 12% from the replacement animals reared off-farm (Table 8.8).

The relative contribution from the wintering period increased with the off-pasture treatments to up to 21% of the total for the slatted floor barn scenario.

Table 8.7. GHG emissions and total carbon footprint of milk ($kg\ CO_2e/kg$ FPCM) from a Waikato dairy farm using different wintering systems

	Grazed Pasture	Stand-off pad	Slatted floor barn
Methane	0.517	0.539	0.550
Nitrous oxide	0.146	0.154	0.157
Carbon dioxide	0.067	0.071	0.074
Other	0.002	0.003	0.003
Total	0.732	0.767	0.784

Table 8.8. Relative contribution of wintering period (70 days), non-wintering period (295 days) and off-farm replacements to the carbon footprint of milk from a Waikato dairy farm using different wintering systems

	Grazed Pasture	Stand-off pad	Slatted floor barn
Non-winter	72%	69%	67%
Wintering	16%	19%	21%
Replacements	12%	12%	12%

Methane was the main GHG emitted at 70-71% of the total carbon footprint, with N_2O at 20% and CO_2 at 9-10% for all scenarios (Table 8.9). Enteric methane dominated total methane emissions while excreta was the main source of N_2O emissions.

Table 8.9. Relative contribution from methane, N_2O and CO_2 to the carbon footprint of milk from a Waikato dairy farm system using different wintering systems

	On Pasture	Stand-off pad	Slatted floor barn
Methane (total)	71%	71%	70%
- enteric CH_4	68%	65%	63%
- dung/FDE/manure CH_4	3%	6%	7%
N_2O (total)	20%	20%	20%
- excreta/FDE/manure N_2O	16%	16%	16%
- N-fertiliser N_2O	3%	3%	3%
CO_2	9%	9%	10%

Table 8.10. Contribution of methane and nitrous oxide emissions ($g\ CO_2e/kg$ FPCM) from excreta produced on the various wintering systems during the wintering period in Waikato

	On Pasture	Stand-off pad	Slatted floor barn
Methane:			
Dung in storage		23.49	35.16
Dung on field	1.07	0.61	
Total	1.07	24.10	35.16
Nitrous oxide:			
Urine on field	8.01	5.34	
Dung on field	6.03	1.07	
Manure in storage		4.20	7.25
Manure application		12.12	16.86
Total	14.04	22.74	24.11

Table 8.11. GHG emissions and total carbon footprint of milk ($kg\ CO_2e/kg$ FPCM) from a Southland dairy farm using different wintering systems

	Grazed Pasture	Solid floor barn	Slatted floor barn
Methane	0.506	0.545	0.545
Nitrous oxide	0.142	0.148	0.149
Carbon dioxide	0.059	0.064	0.066
Other	0.002	0.002	0.002
Total	0.709	0.759	0.762

Enteric methane emissions per kg FPCM were the same for all scenarios but the contribution from dung/manure over winter was 33-fold higher for the slatted-floor barn system than for the all-grazing scenario (Table 8.10). During the winter period, N_2O emissions from manure associated with storage and application from the slatted floor barn scenario were 72% higher than from the all-grazing scenario, with the stand-off pad scenario being 62% higher.

8.3.2. Southland Dairy Farm

The total carbon footprint of milk was lowest for the all-grazing scenario (Table 8.11). The carbon footprint for the slatted floor barn and solid floor barn systems were similar and were 6.9-7.2% higher than for the all-grazing scenario.

The GHG emissions associated with the 70 day wintering period contributed 17% to the total carbon footprint for the Southland all-grazing scenario, with 71% coming from the remainder of the year (295 days including lactation period) and 12% from the replacement animals off-farm (Table 8.12). The relative contribution from the wintering period increased to 23% of the total for the off-pasture scenarios.

Table 8.12. Relative contribution of wintering period, non-wintering period and off-farm replacements to the carbon footprint of milk from a Southland dairy farm using different wintering systems

	Grazed Pasture	Solid floor barn	Slatted floor barn
Non-winter	71%	66%	66%
Wintering	17%	23%	23%
Replacements	12%	11%	11%

Table 8.13. Relative contribution from methane, N_2O and CO_2 to the carbon footprint of milk from a Southland dairy farm system using different wintering systems

	On Pasture	Solid floor barn	Slatted floor barn
Methane (total)	71%	72%	72%
- enteric CH_4	69%	64%	64%
- dung/FDE/manure CH_4	3%	8%	8%
N_2O (total)	20%	19%	19%
- excreta/FDE/manure N_2O	16%	16%	16%
- N-fertiliser N_2O	3%	3%	3%
CO_2	8%	9%	9%

Methane was the largest GHG emission at 71-72% of the total carbon footprint, with N_2O at 19-20% and CO_2 at 8-9% for all scenarios (Table 8.13). Enteric methane constituted 89-96% of the total methane emissions, while dung, FDE and manure contributed 4-11%. For N_2O emissions, the main contributor was excreta, FDE and manure at 81-82%, with N fertiliser contributing 15-16%.

Methane emissions from excreta collected from the solid and slatted floor barn scenarios were the same at 32-fold higher than from dung deposited on pasture in the all-grazing scenario (Table 8.14). Likewise, excreta-related N_2O emissions were 29-30% higher than from the all-grazing scenario.

Table 8.14. Contribution of methane and nitrous oxide emissions ($g\ CO_2e/kg$ FPCM) from animal excreta from the various wintering systems during the wintering period in Southland

	On Pasture	Solid floor barn	Slatted floor barn
Methane			
Dung in storage		39.18	39.18
Dung on field	1.23		
Total	1.23	39.18	39.18
N_2O			
Urine on field	13.96		
Dung on field	3.22		
Manure in storage		6.52	6.67
Manure application		15.87	15.49
Total	17.18	22.39	22.16

Table 8.15. Scenario analysis of the effects of changing the type of supplementary feed that is fed to dairy cows over winter in the slatted floor barn system on the carbon footprint of milk

	Waikato			Southland		
	Pasture silage	Barley grain	Palm kernel expeller	Pasture silage	Barley grain	Palm kernel expeller
Extra feed (kg DM/ha)	1771	1771	1771	2417	2417	2417
g CO_2e/kg FPCM from extra feed	9.1	42.6	144	14.0	58.1	197
Carbon footprint of milk (g CO_2e/kg FPCM)	784	818	919	762	806	945
% rel. to pasture silage		+4%	+17%		+6%	+24%

Table 8.16. Effect of cow wintering system on annual farm N and P losses and on multiple environmental impact indicators

	Waikato			Southland		
	Grazing	Stand-off pad	Slatted floor barn	Grazing	Slatted floor barn	Solid floor barn
On-farm emissions:						
N leaching kg N/ha	33	29	25	28	22	22
Ammonia kg N/ha	47.8	62.0	73.0	45.1	68.6	67.7
P runoff kg P/ha	0.47	0.50	0.50	0.66	0.68	0.68
Impact Category (per kg FPCM):						
Carbon footprint kg CO_2e	0.732	0.767	0.784	0.705	0.757	0.753
Aquatic eutrophication potential[1] g PO_4e	2.92	3.20	3.42	2.67	3.35	3.14
Acidification potential[2] mmol H^+e	13.9	17.5	20.2	12.8	18.6	18.4
Particulate matter[2] g $PM_{2.5}e$	0.317	0.396	0.456	0.291	0.419	0.413
Photochemical ozone formation[2] g NMVOCe	0.563	0.613	0.626	0.504	0.566	0.564

[1] Using CML v3.03 (accounts for N leached, NH_3 and P losses).
[2] Using ILCD 2011 v1.08.

8.3.3. Feed Sensitivity Analysis

A sensitivity analysis of the effects of changing the type of supplementary feed that was fed to dairy cows over winter in the slatted floor barn system illustrated that the carbon footprint of milk could be increased by up to 24% from the use of palm kernel expeller (PKE) instead of pasture silage (Table 8.15). In contrast, there was only a small increase (up to +6%) when barley grain was assessed as an alternative supplementary feed source.

8.3.4. Other Environmental Impact Indicators

It is important to also examine potential effects of the different wintering systems on environmental indicators other than climate change. On the dairy farm, the use of the stand-off pad or housing facilities were estimated to decrease N leaching losses (using the OVERSEER model; Wheeler et al. 2003) due to less urine-N deposition during the high loss risk period of winter (Table 8.16). The reduction was greater for the housed barn systems than the stand-off, since some winter grazing still occurred with the stand-off system. However, ammonia emissions increased with the housing systems, associated with the manure storage and application. A very small increase in P runoff was estimated (using OVERSEER) from the manure applied to land.

Aquatic eutrophication potential was estimated to increase by up to 25% from the slatted floor barn system compared to winter grazing (Table 8.16). This was mainly due to the relatively greater effects from increased ammonia compared to the reduced N leaching. Similarly, the housing and stand-off systems resulted in higher estimates of acidification potential and particulate matter, primarily due to the greater ammonia emissions. A small increase (by up to 12%) in photochemical ozone formation potential with housing was also calculated, associated with NOx from increased fuel use and increased CH_4 emissions from the manure system.

8.4. Discussion

The lowest GHG emissions per kg milk (i.e., the carbon footprint) occurred from the standard all-year grazing scenario. The winter housing scenarios resulted in a 7% increase in the carbon footprint of milk in both the Waikato and Southland systems, while the combination stand-off pad and grazing system was intermediate (+5%). This increase from the housed system was partly due to a small increase in CO_2 emissions associated with infrastructure (from production of the structure and its maintenance) and greater use of fuel for cut-and-carry of pasture to the housed cows over winter. However, most of the increase from the housed system was due to an increase in CH_4 and N_2O emissions associated with the stored effluent from housed cows and from N_2O emissions after subsequent application to land, as found in some other studies reviewed in Chapter 7 (e.g., Hagemann et al. 2011). Of the total increase in carbon footprint from the housing system, 5% was from CO_2 from infrastructure related emissions, 3% from CO_2 from fuel use due to cut-and-carry operations, 3% from CO_2 from fuel and electricity for effluent management, 15% from N_2O from effluent and 73% from CH_4 from stored effluent, on average across the Waikato and Southland farm systems. These results are highly dependent on the CH_4 and N_2O emission factors used for the analysis. Luo et al. (2008) measured N_2O emissions in the field in winter grazing and housed systems in Waikato and found lower total N_2O emissions from manure from the housed system compared to excreta over winter. While the actual emissions will depend on climatic conditions, it must be noted that the current study used a tier 2 methodology which assumes a constant emission factor through all months of the year. If a mechanistic tier 3 model had been used in the current study, it is likely that lower emission rates would have been estimated from effluent applied in spring/summer when soil moisture status is drier than in winter when excreta from grazed pasture is deposited. Nevertheless, most of the increase from housing was due to CH_4 emissions from stored effluent, but there is moderate uncertainty associated with the CH_4 emission factor used.

The overall effects of wintering systems on the annual carbon footprint of milk production was relatively small. This was due to the wintering (70 day) period used for the scenarios only contributing 16-23% to the total carbon footprint, with most of it occurring during the rest of the year.

Life cycle assessment is the preferred methodology for evaluation of different farm system scenarios such as those used in the current study. This is because it accounts for all sources that contribute to GHG emissions. Thus it included emissions associated with the production and maintenance of the infrastructure (housing and yards), energy used in milking and feeding the cows, replacement animals required to maintain a milk-producing herd, and emissions associated with the production and use of feed. The potential effect of the latter factor was illustrated in the sensitivity analysis for the slatted floor scenarios by assessing the effects of providing the winter feed requirements from brought-in barley grain (locally produced) or PKE (from south-east Asia). The carbon footprint of milk increased by 4-6% and 17-24% for barley and PKE, respectively. The higher value for PKE (even after allocation for co-products) is associated with the higher production and transport related emissions from PKE and particularly to emissions from land use change (i.e., due to deforestation for production of palm trees on mineral and peat soils using Blonk 2014) (Ledgard et al. 2016). The relative carbon footprint of milk between housing and partial grazing systems in the study of Wolf et al. (2017) was also influenced by the source of supplementary feeds used.

Normally, an important feature of the use of LCA is evaluation of multiple environmental indicators and not just GHG emissions. This is important because while housing may lead to an increase in GHG emissions it may have benefits in other areas such as in reduced effects on water quality through reduced N leaching, or to negative effects such as from greater ammonia emissions. The latter was associated with higher estimates from housed systems for Eutrophication potential, Acidification potential and Particulate matter, as also found by Arsenault et al. (2009) and Guerci et al. (2013). The greater fuel use from cutting and carrying feed for cow wintering can also affect other indicators including an increase in Photochemical ozone formation potential.

The scenario analyses carried out in the present study and the assumptions used were a simplistic representation of what might happen in practice. Ideally, a wider range of system analyses would have been used including associated use of a farm system model to optimise the system. Optimisation can lead to changes in feed requirements with winter housing, by changing practices such as time of calving, time of drying off and how feed was used in the farm system. The current study also did not include evaluation of some of the other potential benefits from housing such as reduced maintenance feed requirements because of warmer temperatures with housing compared to outdoor grazing. Nevertheless, the current analysis did provide useful information on potential factors that can increase the carbon footprint of milk production from winter housing systems. However, it also highlighted potential options for reducing the carbon footprint from housed systems. These include use of systems to capture and use the CH_4 emitted from stored effluent (which would avoid the main factor that increased the carbon footprint) and the associated energy captured could offset some electricity used for milking and water-heating; and various mitigation options to reduce ammonia and N_2O emissions associated with the storage and application of effluent. It is recommended that more detailed farm system and scenario analyses are carried out to evaluate the full range of factors associated with use of different wintering systems.

CONCLUSION

This model analysis indicated that the use of stand-off pads or housing systems for wintering dairy cows can lead to an increase in the carbon footprint of milk produced from dairy farm systems (e.g., by up to 7%), where there are no other farm system changes. This was due in part to increased use of fossil fuels with cut-and-carry of pasture and establishment and maintenance of infrastructure. However, the main predicted increase in GHG emissions was from management of effluent collected from the pad or housing systems and particularly from methane emissions from stored effluent. If winter feeding during housing is associated with feeds with a relatively high carbon footprint such as PKE, there can be an even larger increase in the carbon footprint of milk produced from dairy farm systems (e.g., by up to 24%). Ideally, the use of stand-off or housing systems need to be integrated with farm system changes and appropriate mitigation practices (e.g., methane capture for energy use; additives to reduce ammonia loss from effluent) to increase overall environmental efficiency and decrease the whole farm system GHG emissions.

ACKNOWLEDGMENTS

The authors acknowledge funding from the Ministry for Primary Industries of New Zealand (MPI) through the Sustainable Land Management and Climate Change (SLMACC) programme.

REFERENCES

Arsenault N., Tyedmers P. and Fredeen A. (2009) Comparing the environmental impacts of pasture-based and confinement-based dairy systems in Nova Scotia (Canada) using life cycle assessment. *International Journal of Agricultural Sustainability* 7: 19-41.

Basset-Mens C., Ledgard S. and Boyes M. (2009) Eco-efficiency of intensification scenarios for milk production in New Zealand. *Ecological Economics* 68: 1615-1625.

Blonk (2014) *Direct land use change assessment tool*. Blonk consultants. Version 2014-1. http://blonkconsultants.nl/en/tools/land-use-change-tool.html.

Chung M. L., Shilton A.N., Guieysse B. and Pratt C. (2013) Questioning the accuracy of greenhouse gas accounting from agricultural waste: A case study. *Journal of Environmental Quality* 42: 654-659.

Clark H. (2001) *Ruminant methane emissions: a review of the methodology used for national inventory estimations*. Report for Ministry of Agriculture and Fisheries, Wellington, New Zealand.

Clark H., Brookes I. and Walcroft A. (2003) *Enteric methane emissions from New Zealand ruminants 1990 and 2001 calculated using an IPCC Tier 2 approach*. Report prepared for the Ministry of Agriculture and Forestry.

Clark D., Malcolm B. and Jacobs J. (2013) Dairying in the Antipodes: recent past, near prospects. *Animal Production Science* 53: 882-893.

DairyNZ (2010) *Facts and figures for New Zealand dairy farmers.* DairyNZ Facts and Figures Version 1.

DairyNZ (2015) *DairyBase*. DairyNZHamilton, New Zealand. https://www.dairynz.co.nz/farm/dairybase/.

DairyNZ (2016) *Pasture growth data*. DairyNZ, Hamilton, New Zealand. https://www.dairynz.co.nz/feed/pasture-management/pasture-growth-data/.

Guerci M., Knudsen M. T., Bava L., Zucali M., Schonbach P. and Kristensen T. (2013) Parameters affecting the environmental impact of a range of dairy farming systems in Denmark, Germany and Italy. *Journal of Cleaner Production,* 54: 133-141.

Hagemann M., Hemme T., Ndambi A., Alqaisi O. and Sultana M. N. (2011) Benchmarking of greenhouse gas emissions of bovine milk production systems for 38 countries. *Animal Feed Science and Technology* 166-167: 46-58.

IDF (2015) International Dairy Federation. A common carbon footprint for the dairy sector: The IDF guide to standard lifecycle assessment. Available at:http://store.fil-idf.org/product/a-common-carbon-footprint-approach-for-the-dairy-sector-the-idf-guide-to-standard-life-cycle-assessment-methodology/.

IPCC (2006) IPCC Guidelines for National Greenhouse Gas Inventories: Volume 4: Agriculture, Forestry and other Land Use. *Intergovernmental Panel on Climate Change.* Paris, France. Available at: http://www.ipcc-nggip.iges.or.jp/public/2006gl/vol4.htm.

IPCC (2007) Climate Change 2007: The Physical Science Basis. *Contribution of Working Group I to the Fourth Assessment Report of the Intergovernmental Panel on Climate Change*. Solomon S, Qin D, Manning M et al. (eds). Cambridge University Press, Cambridge, UK.

Kelliher F. M., de Klein C. A. M., Li Z. and Sherlock R. (2005) *Review of nitrous oxide emission factor (EF3) data.* Report for the Ministry of Agriculture and Forestry, Wellington, New Zealand.

Laubach J., Pratt C., Shilton A., Guieysse B., Chung M., Heubeck S., Park J., Craggs R., Woodward B. and van der Weerden T. (2013) *Review of gaseous emissions of methane, nitrous oxide and ammonia, and nitrate leaching to water, from farm dairy effluent storage and application to land.* Report for the Ministry for Primary Industries, New Zealand.

Ledgard S. F. and Brier G. (2004) *Estimation of the proportion of animal excreta transferred to the farm dairy effluent system.* AgResearch. Report for the Ministry for Agriculture and Forestry, New Zealand.

Ledgard S. F., Basset-Mens C., Boyes M. and Clark H. (2008) *Carbon footprint measurement - Milestone 6: Carbon footprint for a range of milk suppliers in New Zealand.* Report to Fonterra. AgResearch, Hamilton. 41p.

Ledgard S. F., Webby R. and Hawke M. (2003) *Improved estimation of N excretion by grazing animals to improve N_2O emission inventories*. Report to MAF. AgResearch Ruakura. 27p.

Ledgard S. F., Boyes M. and Brentrup F. (2011) Life cycle assessment of local and imported fertilisers used on New Zealand farms. In: *Adding to the knowledge base for the nutrient manager*. (Eds. Currie L.D. and Christensen C.L.). http://flrc.massey.ac.nz/publications.html. Occasional Report No. 24, Fertilizer and Lime Research Centre, Massey University, Palmerston North, New Zealand. 13p.

Ledgard S. F., Chobtang J., Falconer S. and McLaren S. (2016) Life cycle assessment of dairy production systems in Waikato, New Zealand. In: *Integrated nutrient and water management for sustainable farming*. (Eds L. D. Currie and R. Singh). http://flrc.massey.ac.nz/publications.html. Occasional Report No. 29. Fertilizer and Lime Research Centre, Massey University, Palmerston North, New Zealand. 8p.

LIC (2011) *NZ dairy statistics 2010-2011.* Livestock Improvement Corporation limited, Hamilton, New Zealand. http://www.lic.co.nz/pdf/DAIRY%20STATISTICS%2010-11-WEB.pdf.

Luo J. and Longhurst B. (2008) *Improved livestock management systems and nutrient losses from manure deposition and treatment – A support document for Overseer nutrient budget model.* Report to AgResearch.

Luo J., Ledgard S. F. and Lindsey S. B. (2008) A test of a winter farm management option for mitigating nitrous oxide emissions from a dairy farm. *Soil Use and Management* 24: 121-130.

MAF (2011) *Waikato/Bay of Plenty Intensive sheep and beef farm: Pastoral monitoring 2011.* Report from Ministry of Agriculture and Forestry, Wellington, New Zealand. 8p.

MfE (2016) *New Zealand's Greenhouse Gas Inventory 1990–2014*, Ministry for the Environment. Wellington, New Zealand. 519p.

Schmidt J. H. (2007) *Life cycle assessment of rapeseed oil and palm oil.* (Ph.D. thesis). Aalborg University, Denmark.

Sinclair A. G., Edmeades D. C. and Ledgard S. F. (1993) The influence of phosphorus, sulphur and nitrogen fertilisers on acidification in pastoral soils. *Proceedings of the NZ Fertiliser Manufacturers' Research Association Conference* 22: 106-116.

Thomas S. M., Ledgard S. F. and Francis G. S (2005) Improving estimates of nitrate leaching for quantifying New Zealand's indirect nitrous oxide emissions. *Nutrient Cycling in Agroecosystems* 73: 213-226.

Wheeler D. M., Ledgard S. F., de Klein C. A. M., Monaghan R. M., Carey P. L., McDowell R. W. and Johns K. L. (2003) OVERSEER® nutrient budgets – moving towards on-farm resource accounting. *Proceedings of the New Zealand Grassland Association* 65: 191-194.

Wolf P., Groen E. A., Berg W., Prochnow A., Bokkers E. A. M., Heijungs R. and de Boer I. J. M. (2017) Assessing greenhouse gas emissions of milk production: which parameters are essential? *International Journal of Life Cycle Assessment* 22: 441-455.

In: Greenhouse Gas Emissions …
Editors: Jiafa Luo and Yan Li
ISBN: 978-1-53611-100-2

Chapter 9

A REVIEW OF THE COST-EFFECTIVENESS OF GRAZING-OFF PRACTICES

Femi Olubode-Awosola, James Turner* and Akiko Horita
AgResearch, Hamilton, New Zealand

9.1. INTRODUCTION

Cost-effectiveness analysis (CEA) compares the relative costs and outcomes (or effects) of alternative courses of action. For example, Environment Bay of Plenty (2007) used CEA to estimate the cost-effectiveness of N leaching mitigation practices. This was expressed as changes in farm profitability, per ha per kilogram (kg) of N leaching ($/kg N). In the case of this study, we are interested in the dollar cost per t-CO_2-eq of GHG emissions reduced by alternative grazing-off technologies. CEA is used to identify a course of action that is cost-effective, in terms of the outcomes sought. There is limited literature on the cost-effectiveness of off-grazing practices in New Zealand, especially dairy housing, as New Zealand is predominantly characterised by pastoral grazing practices.

Here we review the cost-effectiveness of wintering-off technologies in dairy farming, in terms of type of housing and length of grazing-off during

* Corresponding Author: Dr James Turner, AgResearch, 10 Bisley Road, Private Bag 3123, Hamilton 3214, New Zealand. Email: James.Turner@agresearch.co.nz.

winter. The review will reveal which variables are important in dairy grazing-off, in terms of GHG mitigation and the cost-effectiveness of different animal confinement systems.

Winter grazing management practices include winter on-farm grazing, winter off-farm grazing, use of winter feed pads, use of stand-off pads, use of wintering pads or combinations (feed and stand-off pad) and use of Herdhome® Shelters (Chapter 2). In this review, we attempt to answer the following questions; what are the capital costs associated with animal confinement systems? What items constitute significant proportions of the total cost of housing systems? What are the costs that vary with length of grazing-off, in terms of hours spent in confinement?

9.2. Cost Structure

In this section, we present information on the different costs associated with using animal housing. In many European countries, it is common practice to house animals throughout the year, rather than graze them on pasture during some seasons. Animal housing involves significant capital investment; the cost of building housing, maintenance, effluent and feed management systems. Brock and Barham (2009) compared the structure and economic performance of different Wisconsin dairy farm management systems, varying the confinement level and associated pasture management systems. It was found that confined systems come with high investment costs, of associated confinement infrastructure. These include feeding and manure management facilities such as automatic scrapers or tractor mounted scrapers.

Herd size influences capital costs, through economies of scale, where the capital cost per animal declines with increasing herd size (Auernhammer 1987; Næss and Stokstad 2011). Næss and Stokstad (2011) studied 44 farms between 1999 and 2006 in Norway to investigate the importance of economies of scale in building costs of barns, compared to other sources of variation in costs. Their analysis showed that construction costs per m^2 decreased with size (area) of barns, up to approximately 1,250 m^2. Similarly, mechanisation costs and total building costs per m^2 decreased up to approximately 1,000 m^2. A further increase in barn area had only limited effect on the building cost per m^2. Pereira et al. (2003) developed a spreadsheet to enable dairy managers to predict the construction costs per cow of alternative animal house designs when considering increasing herd size. With a case study of Galicia in Spain, the authors compared six of the common dairy housing designs and showed that a range of herd sizes could have similar costs in terms of choice of design, whereas designs for smaller herd sizes have higher housing costs per cow. Auernhammer (1987) reported that increasing the efficiency of a dairy house through increasing herd size is an important means of reducing the capital cost of dairying housing in Germany. The design of animal housing, by modifying existing tie-stalls barns into cubicles, is cost-effective only when milk yield per cow and herd size exceed minimum levels.

Barbari et al. (2007) calculated and compared the costs of building, equipment, labour, electricity usage, machines and litter for nine standard cowsheds. Each shed had differing lying areas and manure management in loose housing systems (where animals were kept loose except during milking and at the time of treatment), for dairy cows in Emilia-Romagna, Italy. The building design that had the lowest costs for manure management featured cubicles with litter and cleaning with automatic scrapers or tractor mounted scrapers.

In summary, capital costs associated with investing in building animal housing are significant. The capital cost is affected by the design, materials and labour used in the construction and the feeding and manure management facilities. However, the capital cost per head can be reduced through economies of scale achieved with larger herds and hence larger housing.

9.3. Costs That Vary with Level of Confinement, Feed and Effluent Management

In this section, we present information about the cost components that vary with level of confinement and different feed and effluent management practices, in animal housing. These costs differ from capital costs in that they vary with changes in animal housing practices.

Besides housing size, design and facilities being the most significant cost items of animal housing, other costs that vary in animal housing include different feed/grazing management systems, length of wintering-off period, and effluent management options within a particular housing system. The overall cost-effectiveness in terms of cost per kilogram of reduced N (kg N) leaching for the winter grazing options in the Environment Bay of Plenty (2007) is presented in Table 9.1. Option 8, Herdhome® full-time was the most cost-effective. This option was one where all cows were off pasture and in the

Herdhome® shelter for 20 hours per day during winter and on the pasture for four hours; during lactation all cows were in the Herdhome® for 12 hours per day and on the pasture for eight hours. The option that had the highest cost per kg N mitigated was Option 1 where all cows were on pasture all year. The Herdhome® full-time had an annual cost of $25.60/kg N reduction in nitrate leached.

Table 9.1. Cost-effectiveness ranking of grazing management options in Rotorua Lake Catchment

Winter grazing option	Grazing regime summary	Cost-effectiveness ($/kg N)
1	All cows on pasture all year	-
2	All cows grazed off farm for 45 days from 1 June	-
3	All cows off pasture and fed on the feed pad for four hours per day during winter and for one hour per day during lactation	418.10
4	All cows off pasture and on the stand-off pad for 12 hours per day during winter	81.30
5	All cows off pasture and on the feed pad for four hours per day and on the stand-off for 12 hours per day during winter During lactation the cows will be on the feed pad for one hour per day	61.70
6	All cows off pasture and in the Herdhome® Shelter for 20 hours per day and for four hours per day on pasture during winter only	56.60
7	All cows off pasture in the Herdhome® Shelter for 20 hours per day during winter and four hours on pasture. During lactation all cows in Herdhome® for four hours per day and 16 hours on pasture	38.10
8	All cows off pasture and in the Herdhome® Shelter for 20 hours per day during winter and for four hours on pasture. During lactation all cows in Herdhome® for 12 hours per day and eight hours on pasture	25.60

Source: Environment Bay of Plenty (2007).

De Klein (2001) used data from an average New Zealand dairy farm to show the economic implications of nil and restricted grazing systems. Increased capital and operating costs were associated with nil and restricted grazing in New Zealand. Based on the average New Zealand dairy farm, the cost/benefit analysis of the nil grazing system suggested a negative return on capital (profit that is gained from capital investment) of about -10%. However, for the restricted grazing system, the average return on capital was about 9% (range: -4 to 25%). This can improve up to 17% (range: 2 to 50%) where an effluent application system or a feed pad is already in place.

Literature outside New Zealand also shows important costs that vary with management practices in animal housing. Rotz et al. (2006) simulated the effects of four nutrient management technologies - ammonia emission barn floor (Barn floor designed to reduce the rate of emissions of ammonia. Emissions of ammonia from the solid part of a barn floor can be reduced by using a smoothly finished, inclined or convex surface), enclosed manure storage, manure injection into the soil, and intra seeding of a grass cover crop on corn land - in terms of reduced N loss and profitability of the technologies on American dairy farms. The most cost-effective technique was a deep injection of manure when a cropping and tillage system was used that allowed this practice. The added equipment and operating cost of about $14/cow per year may have been returned in the fertilizer value of the nitrate saved. It also had the advantage of odour reduction and the retention of other nutrients. For the simulated 100-cow farm, the predicted N saving through reduced leaching and denitrification losses was 8 kg N/ha. The annual fertilizer value of this N was about $7.80/ha or $8.00/cow.

Rotz et al. (2005) reviewed how manure handling techniques (cover crops, low emission barns, covered manure storages and direct injection of manure into the soil) being trialled on pilot experimental farms to reduce N loss are being applied to commercial farms in North America and Europe. The review reported that these techniques can help farmers to increase profitability and reduce N loss; however, the techniques created net costs to the farmers especially the manure handling costs of $190/cow, resulting in a loss in the annual net return to farm management. Sedorovich et al. (2007) simulated the effect of manure handling and tillage systems on P loss from farms in Pennsylvania. The simulation showed that on a 100-cow dairy farm, a manure handling strategy that used a 6-month storage and application by injection decreased total P loss by 19% compared to daily surface application, but the annual farm net return was decreased by $57/cow.

In summary, the level of confinement, and type of feed and effluent management practices have an important influence on the costs associated with animal housing. Each of them, alone or in combination with other practices, can be used to increase the overall economic efficiency of animal housing. In addition, application of each of these practices increases the efficiency of use of farm inputs particularly fertilizer, feed and labour.

CONCLUSION

There is little information in the literature about cost-effectiveness of grazing-off pasture using animal housing in New Zealand as pastoral grazing predominates. Specifically, there are few comparisons across housing types or confinement lengths. Comparisons of different grazing/feed management systems have usually been within a particular housing system with some reference to the costs of the housing structure. From the review, the most significant costs are the building (or capital cost), and costs associated with feed and effluent handling. Costs that vary with operation level (e.g., herd size) include labour costs associated with manure and animal handling. From information from outside New Zealand, grazing-off animals in animal housing involves high capital costs. These are often managed through economies of scale from reduced cost per head associated with a larger herd size and hence a larger housing structure. Level of confinement, and feed and effluent management practices, can also be varied to reduce the overall cost of animal housing while also increasing the efficient use of farm inputs such as fertilizer, labour and feed.

ACKNOWLEDGMENTS

The authors acknowledge funding from the Ministry for Primary Industries of New Zealand (MPI) through the Sustainable Land Management and Climate Change (SLMACC) programme. Penny Payne kindly provided additional English editing for this chapter.

REFERENCES

Auernhammer H. (1987) Economic dairy cattle housing can only be achieved on big farms and by rebuilding. *Milchpraxis* 25: 120-122.

Barbari M., Ferrari P. and Rossi P. (2007) *Technical and economical analysis of manure management in loose housing systems for dairy cows*. In the Proceedings of the Sixth International Dairy Housing Conference Proceeding, 16-18 June 2007, Minneapolis, Minnesota, USA, Publication Date 16 June 2007. ASABE Publication Number 701P0507e.

Brock C. and Barham B. (2009) Farm structural change of a different kind: Alternative dairy farms in Wisconsin - Graziers, organic and Amish. *Renewable Agriculture and Food Systems* 24: 25-37.

De Klein C.A.M. (2001) Analysis of environmental and economic implications of nil- and restricted-grazing systems designed to reduce nitrate leaching from New Zealand dairy farms. II. Pasture production and cost/benefit analysis. *New Zealand Journal of Agricultural Research* 44: 217-235.

Environment Bay of Plenty (2007) *Cost-effectiveness Analysis of Grazing.*

Næss G. and Stokstad G. (2011) Dairy barn layout and construction: Effects on initial building costs. *Biosystems Engineering* 109: 196-202.

Pereira J.M., Álvarez C.J. and Barrasa M. (2003) Prediction of dairy housing construction costs. *Journal of Dairy Science* 86: 3536-3541.

Rotz C.A., Taube F., Russelle M.P., Oenema J., Sanderson M.A. and Wachendorf M. (2005) Whole-farm perspectives of nutrient flows in grassland agriculture. *Crop Science* 45: 2139-2159.

Rotz C.A., Oenema J. and van Keulen H. (2006) Whole farm management to reduce nutrient losses from dairy farms: A simulation study. *Applied Engineering in Agriculture* 22: 773-784.

Sedorovich D.M., Rotz C.A., Vadas P.A. and Harmel R.D. (2007) Simulating management effects on phosphorus loss from farming systems. *American Society of Agricultural and Biological Engineers* 50: 1443-1453.

In: Greenhouse Gas Emissions …
Editors: Jiafa Luo and Yan Li
ISBN: 978-1-53611-100-2

Chapter 10

HOUSED VERSUS RESTRICTED WINTER GRAZING SYSTEMS: A COST ANALYSIS IN SOUTHLAND AND WAIKATO, NEW ZEALAND

Femi Olubode-Awosola, James Turner* and Akiko Horita
AgResearch, Hamilton, New Zealand

10.1. INTRODUCTION

In Chapter 9, we reviewed the literature regarding the cost effectiveness of two grazing-off strategies in dairy farming; providing housing and varying the length of time grazing-off. The review showed that the most significant costs are the housing (capital) costs, feed and effluent management, and the associated labour costs. In this chapter, we demonstrate how the different grazing-off strategies compare in terms of cost structure and greenhouse gas (GHG) emissions, using case studies of typical Waikato and Southland farms.

* Corresponding Author: Dr James Turner, AgResearch, 10 Bisley Road, Private Bag 3123, Hamilton 3214, New Zealand. Email: James.Turner@agresearch.co.nz.

10.2. Case Study Farms and Methods

In this section, we start by presenting information about two case studies of hypothetical grazing-off strategies - one for Waikato and the other for Southland. The parameters of the case study farms and the assumptions used in estimating the costs associated with the different strategies are also presented.

10.2.1. Case Study Farms

The Waikato and Southland case study farms are the same farms used for the study of total GHG emissions (Chapter 8). The case study farms were representative of average 2010/11 season farm data, based on the Dairybase® survey for the Waikato and Southland regions. The Dairybase® data was adjusted for the regional average using the Livestock Improvement Corporation (LIC) production data. The cost items that differ between the wintering strategies for the case study farms are presented in Table 10.1 and Table 10.2 for the Waikato and Southland case study farms respectively. The GHG emissions from the different strategies are also presented in the tables below (as from Chapter 8). The differences in the strategies in terms of confinement types, length of confinement, feed management, and animal and effluent handling were the bases for estimates of the GHG emissions. They are recorded in metric tonne carbon dioxide equivalent (t CO_2-eq), per cow, for these case study farms.

Table 10.1. Wintering strategies and GHG emissions for the Waikato case study farm

Winter-off period (weeks)	0	10	10
Wintering system	On Pasture	cut & carry + slatted floor barn	8hrs graze + 16hrs stand-off
Average area (effective ha)	110	110	110
Fertiliser (kg N per on-farm ha)	126	126	126
Supplement-Pasture silage (kg DM/ha)	1417	1417	1417
GHG emission (t CO_2-eq per ha)	10.80	11.58	11.32
GHG emission (t CO_2-eq per cow)	3.75	4.02	3.93

Data in this table was extracted from Chapter 8.

Table 10.2. Wintering strategies and GHG emissions for the Southland case study farm

Stand-off period (weeks)	0	10	10
Wintering system	On Pasture	Cut & carry + slatted floor barn	Cut & carry + solid floor barn
Average area, (effective ha)	206	206	206
Fertiliser (kg N per on-farm ha)	112	112	112
Supplement - Pasture silage kg (DM/ha)	1957	1957	1957
GHG emission (t CO_2-eq per ha)	10.60	11.36	11.33
GHG emission (t CO_2-eq per cow)	4.03	4.32	4.31

Data in this table was extracted from Chapter 8.

The Waikato case study farm could either winter 316 cows on pasture (110 ha effective area) or follow any of the assumed two grazing–off strategies, as follows:

1) On-farm wintering on pasture (on pasture)
2) Cut and carry pasture to feed cows, confine cows in slatted floor barn for 10 weeks (cut & carry + slatted floor barn)
3) Grazing cows for 8 hours/day, confine cows on stand-off pad for 16 hours/day with feed pad for 10 weeks (8 hrs grazing + 16 hrs stand-off pad + feed pad)

In the Southland case study farm, a typical farm of 206 ha effective area could winter 542 dairy cows. In terms of wintering strategies, this farm could either winter on the farm on pasture, or follow any of the assumed two wintering strategies. Therefore, the three wintering strategies considered are:

1) On-farm wintering on pasture (on pasture)
2) Cut and carry pasture to feed cows, confine cows in slatted floor barn for 10 weeks (cut & carry + slatted floor barn)
3) Cut and carry pasture to feed cows, confine cows in Solid floor barn for 10 weeks (cut & carry + solid floor barn)

Table 10.3. Assumptions and parameter estimates for the case study farms

Sn	Cost items and assumptions	Estimate	Source
	Operating costs		
1	Cost of cut and carry surplus grass silage to platform ($/kg DM)	0.04	Beukes et al. (2011)
2	Cost of pasture silage supplement ($/kg DM)	0.212	DairyNZ (2010) Lincoln University (2012)
3	Cost of emptying effluent bunkers in slatted floor barn ($/cow/year)	20	Herdhome® (2009, 2013)
4**	Cost of handling effluent from the 16hr stand-off pad assumed at 2/3 of that from the slatted floor barn ($/cow/year)	13	Own assumption
5	Cost of bedding materials in stand-off pad and Deep litter barn ($/cow/year)	40.00	Market quoted 2012
	Fixed costs		
6	Interest* on capital costs of slatted floor barn in Waikato case farm ($/cow/year)	118	Herdhome® (2009, 2013) Dexcel (2005)
7	Interest* on capital costs of stand-off in Waikato case farm ($/cow/year)	35	Redpath (2013)
8	Interest* on capital costs of slatted floor barn in Southland Case farm ($/cow/year)	142	Beukes et al. (2011)
9	Interest* on capital costs of Deep litter barn ($/cow/year)	33.6	Redpath (2013)
10	Interest* on capital costs of feed-pad ($/cow/year)	10	DairyNZ (2008)
11	Interest* on capital cost of silage bunkers ($/cow)	16.54	Herdhome® (2009, 2013)
12	Cost of annual maintenance of slatted floor barn ($/cow/year)	10	Herdhome® (2009, 2013)
13	Cost of maintenance of stand-off pads ($/cow/year) at 9% of cost of installation	3.02	Herdhome® (2009, 2013)
	Other Cost		
14	Cost of N fertilizer - Urea ($/tonne)	860	Lincoln University (2012)

* Interest rate assumed at 8%.

** This was based on the assumption that cows will be confined for two thirds of the time that they would be confined in the slatted floor barn.

10.2.2. Methods: Assumptions and Estimation Approach

In estimating the costs associated with the different grazing-off strategies for the case study farms, we used the information about the farm in terms of the number of cows and the effective farm area. In addition, we used only the parameters that differ between the wintering strategies in terms of housing types, length of time off the pasture, feed, and animal effluent management systems. For each of the case study farms, it was assumed that the milking cows were confined in some way on the farm for 10 weeks during the winter period. The parameters that vary by grazing-off strategies include:

- pasture silage supplement (kg DM intake per ha)
- amount of on-farm pasture to harvest – specifically to cut and carry to animals in the confinement (either slatted floor barn, deep litter barn or stand-off pad)

Based on the information that differs between the strategies, a number of cost assumptions were made, as presented in Table 10.3.

The cost estimates were related to the GHG emissions and are summarized in Figures 10.1 and 10.2 for the Waikato and Southland case study farms, respectively, in the Results section.

10.3. Results

10.3.1. Costs Associated with Wintering Strategies in the Waikato Case Study Farm

From Table 10.4, the operating costs that vary among these strategies include feed (labour associated with cut and carry of pasture to feed cows in the slatted floor barn; off-farm pasture silage supplement), effluent management labour costs and bedding costs. Although the GHG emissions associated with using bedding in a stand-off pad was not considered in the estimated GHG emissions, bedding was considered a critical cost in stand-off pads (DairyNZ 2003; Dexcel 2005; DairyNZ 2012). As such, we have considered this cost in the case studies. The fixed costs that vary among the strategies include the cost of animal housing (interest on capital costs), cost of repair and maintenance of the housing, and equipment and facilities associated with housing the animals.

The on-farm wintering system has the lowest operating cost at approximately $100 per cow, per year. The operating cost associated with the cut & carry + slatted floor barn strategy is approximately $140 per cow, per year. The operating cost associated with 8 hours grazing + 16 hours stand-off + feed pad is approximately $160 per cow, per year. The fixed (capital) cost is zero for the on-farm strategy. It is approximately $40 per cow, per year for the 8 hours grazing + 16 hours stand-off pad + feed pad strategy and $145 per cow, per year for the cut & carry + slatted floor barn strategy.

Table 10.4. Costs associated with wintering strategies for the Waikato case study farm

	Wintering strategies		
Costs associated with wintering strategies	On Pasture	cut & carry + slatted floor barn	8hrs grazing + 16hrs stand-off pad + Feed pad
Operating Costs			
Feed			
On-farm cut & carry (including labour) ($/cow)			
Off-farm pasture silage ($/cow)		14	
Labour	104	104	104
Effluent Management ($/cow)		20	13
Bedding ($/cow)			40
Total operating costs ($/cow)	**104**	**138**	**157**
Fixed Costs			
Housing(Buildings)/stand-off pads			
Interest ($/cow)		118	35
Repair and Maintenance ($/cow)		10	3
Equipment/facility ($/cow)		17	10
Total Fixed ($/cow)		**145**	**48**
Other Costs			
Fertilizer costs ($/cow)		38	38
Total Costs ($/cow)	**142**	**321**	**243**
*GHG emission (t CO_2-eq per cow)	3.75	4.02	3.93
Cost per CO_2-eq ($/t CO_2-eq)	**37.87**	**78.85**	**61.83**

* Estimates from Chapter 8.

The cost of fertilizer is approximately $40 per cow, per year for the on-farm system, cut & carry + slatted floor barn, and the 8 hours grazing + 16 hours stand-off pad + feed pad strategies. This means that if the case study farm winters the cows on pasture, the total cost could be approximately $140 per cow, per year. However, if the cut & carry + slatted floor barn strategy is used, the total cost would be approximately $320 per cow, per year. If the 8 hours grazing + 16 hours stand-off pad + feed pad strategy is used, the total cost could be approximately $240 per cow, per year.

In terms of GHG emissions, the wintering-off strategies have somewhat similar amounts of GHG emissions. The GHG emissions were estimated at 3.8t CO_2-eq per cow for the on-farm strategy, 4.0t CO_2-eq per cow for the cut & carry + slatted floor barn strategy, and 3.9t CO_2-eq per cow for the 8 hours grazing + 16 hours stand-off pad + feed pad strategy.

The summary of the costs per tonne of CO_2-eq GHG emission is presented in Figure 10.1. The cost associated with the on-pasture strategy is about $8 per t CO_2-eq. The 8 hours grazing + 16 hours stand-off pad + feed pad costs about $62 per t CO_2-eq. Between these two strategies, the operating costs contribute slightly more of the difference than the capital costs. This is due to the cost of bedding and the cost of labour in managing effluent in the stand-off pad. However, the GHG emissions associated with bedding were not considered in the total GHG emissions. If bedding costs were accounted for, the costs per t CO_2-eq might be more comparable as the GHG emissions will go up and the cost per t CO_2-eq will be smaller than the current estimated $60 per t CO_2-eq for the 8 hours grazing + 16 hours stand-off pad + feed pad strategy.

In summary, the cut & carry + slatted floor barn strategy was inferior, both in terms of GHG emission and the cost per t of CO_2-eq GHG emission, while the on-pasture wintering system was the best in terms of GHG emissions and the cost associated with the level of emission.

10.3.2. Costs Associated with Wintering Strategies in the Southland Case Study Farm

As for the Waikato case study farm, the operating costs that vary among these wintering strategies include feed, effluent management, labour and bedding costs (Table 10.5). The fixed costs that vary include the cost of animal housing; in terms of interest on capital cost, cost of repair and maintenance of the housing, as well as equipment and facilities associated with housing the animals.

There are some relative differences in operating costs among the strategies. The on-farm wintering strategy has the lowest cost, at approximately $158 per cow, per year. The slight difference means little trade-off between these strategies. The cost of bedding makes the operating cost higher in the cut & carry + solid floor barn strategy, compared to the cut & carry + slatted floor barn.

The fixed cost (from capital investment) is zero for the on-farm wintering strategy. The fixed costs are same for the slatted floor and solid floor barn strategies. Although fixed costs are higher in the slatted floor barn strategy than in the solid floor barn strategy, the difference is mainly the capital cost of the slatted floor barn, which is higher than for the solid floor barn. This means that if the Southland case study farm uses grazing-off strategies, the fixed cost could range from approximately $50 per cow, per year in the solid floor barn strategy, to approximately $170 per cow, per year in the slatted floor barn strategy.

In terms of total costs, on-pasture wintering has the lowest cost. This is approximately two-thirds of the costs in the cut & carry + solid floor barn strategy and approximately half of the cost in the cut & carry + slatted floor barn. This suggests that the cost of pasture silage supplement is slightly higher than the cost of applying fertilizer and spending on labour, to cut and carry.

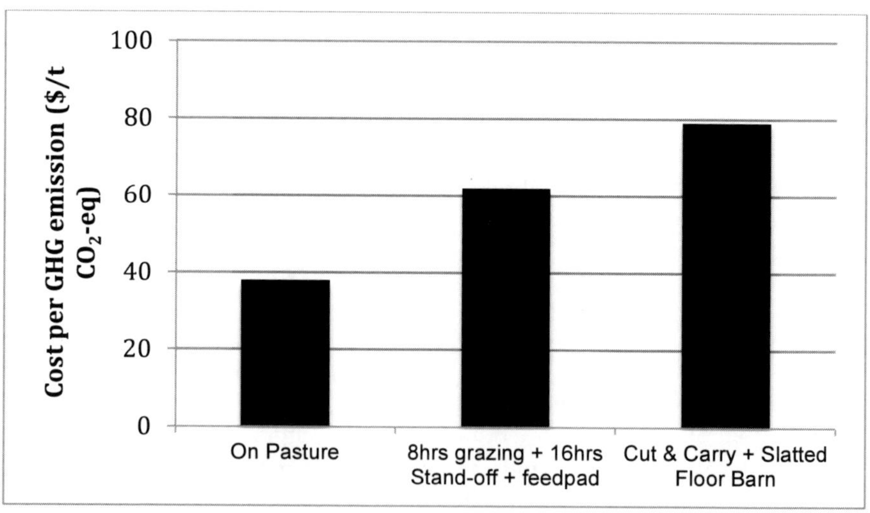

Figure 10.1. Cost associated with GHG emission by wintering systems in the Waikato case study farm.

Table 10.5. Costs associated with wintering strategies for the Southland case study farm

	Wintering strategies		
Costs associated with wintering strategies	On Pasture	cut & carry + slatted floor barn	cut & carry + solid floor barn
Operating Costs			
Feed			
On-farm cut & carry (including labour) ($/cow)		8	8
Off-farm pasture silage ($/cow)	158	158	158
Labour			
Effluent Management ($/cow)		20	20
Bedding ($/cow)			40
Total operating costs ($/cow)	**158**	**186**	**226**
Fixed Costs			
Housing(Buildings)/stand-off pads			
Interest ($/cow)	-	142	37
Repair and Maintenance ($/cow)	-	10	3
Equipment/facility ($/cow)	-	17	10
Total Fixed ($/cow)	-	**169**	**50**
Other Costs			
Fertilizer costs ($/cow)	37	37	37
Total Costs ($/cow)	**195**	**392**	**313**
*GHG emission (t CO_2-eq per cow)	4.03	4.32	4.31
Cost per CO_2-eq ($/t CO_2-eq)	**48.39**	**90.74**	**72.62**

* Estimates from Chapter 8.

In terms of the GHG emissions, the GHG emissions from the cut & carry + slatted floor barn and the cut & carry + solid floor barn strategies are comparable. The emissions are lowest in the on-pasture wintering strategy, at about 4.03t CO_2-eq, but highest in the cut & carry slatted floor barn strategy, at about 4.32t CO_2-eq.

The summary of the costs associated with the GHG emissions is presented in Figure 10.2. The on-pasture strategy is lowest, at approximately $48 per t CO_2-eq. The cost is highest in the cut & carry slatted floor barn strategy, at approximately $91 per t CO_2-eq.

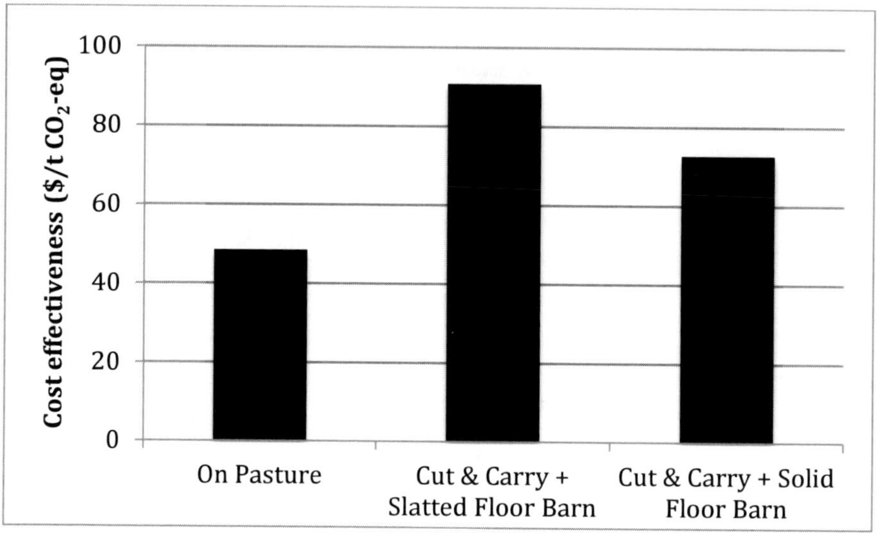

Figure 10.2. Cost associated with GHG emission by wintering systems in the Southland case study farm.

In summary, comparing the costs associated with the GHG emissions between the cut & carry + solid floor barn and the cut & carry + slatted floor barn strategies suggests that the cut & carry + solid floor barn strategy is cheaper.

Conclusion

The GHG emissions associated with the capital and infrastructure items only make a small contribution to the total GHG emissions. In both case studies, wintering on pasture is least expensive. Therefore, in terms of costs, the capital cost of housing is the main component of the costs associated with wintering strategies. Operating costs are an additional important cost.

In conclusion, the results give an indication of how the systems compare in terms of costs and reduction in GHG emissions.

- The capital cost of housing is the main component of the costs associated with wintering strategies.
- Stand-off pads are cheaper than deep litter barns, which are cheaper than slatted floor barns.

- Other costs that are important are the operational costs associated with feed and effluent management.
- Generally, wintering on pasture is more cost effective per t of CO_2-eq GHG emission, due to lower fixed and operating costs per cow and lower GHG emissions per cow.

This study identified that wintering the cows on pasture all year is the most cost effective option for managing GHG emissions. Interestingly, this wintering strategy was identified as having the highest mitigation cost per kg of N, in study by Environment Bay of Plenty (2007).

These case studies have allowed us to provide insights on how the level of confinement, feed, labour and effluent management can be varied to reduce the overall cost of wintering cows in housing. From the literature review we identified that the high cost of capital in animal housing can be managed (reduced) through economies of scale, as reduced capital cost per head can be realised from a larger herd size, and hence a larger housing structure. Assuming constant herd and housing size is therefore one limitation of the two case studies featured in this chapter. In addition, there are potential positive and negative animal welfare and environmental effects from wintering-off animals in animal housing that we have not considered in these case studies.

ACKNOWLEDGMENTS

The authors acknowledge funding from the Ministry for Primary Industries of New Zealand (MPI) through the Sustainable Land Management and Climate Change (SLMACC) programme. Penny Payne kindly provided additional English editing for this chapter.

REFERENCES

Beukes P.C., Gregorini P., Romera A.J. and Dalley D.E. (2011) The profitability and risk of dairy cow wintering strategies in the Southland region of New Zealand. *Agricultural Systems* 104: 541-550.

DairyNZ (2003) *Farm for tomorrow - Standing off - A cow's perspective.* DairyNZ brochure.

DairyNZ (2008) *feed pads*. DairyNZ farmfact 8-1.

DairyNZ (2010) *Facts and figures for New Zealand dairy farmers*. DairyNZ Facts and Figures Version 1.

Dexcel (2005) *Minimising muck, maximising money: Stand-off and feed pad design and management guidelines*. Dexcel Ltd, Hamilton.

Environment Bay of Plenty (2007) *Cost-effectiveness Analysis of Grazing.*

Herdhome® (2009) *North Waikato Case Study 4 - From a feed pad and spring calving to Herdhome®s shelters and split calving.*

Herdhome® (2013) *From a wood chip based stand-off pad to Herdhome® shelters Once a day milking (OAD). Case study 5 - Bay of Plenty.* At http://www.Herdhome®s.co.nz/portal/index.php?option=com_content&task=view&id=70&Itemid=82.

Lincoln University (2012) *Financial Budget Manual 2012/13*, Lincoln University, Christchurch, New Zealand.

Redpath (2013) *Clear-roofed, deep litter stand-off shelters*, Redpath Brochure.

ABOUT THE EDITORS

Dr *Jiafa Luo*, a senior scientist at AgResearch New Zealand, has established himself as a key science leader in the agricultural nutrient management and greenhouse gas research areas. He has led a number of important multi-organizational research programmes that operate throughout New Zealand and internationally. Dr Luo has published over 100 papers and developed a strong international reputation, as is evidenced by his many invitations to attend international workshops, conferences and other forums. He is an editorial board member for the journal "Agriculture, Ecosystems and Environment" and for "Communications in Soil Science and Plant Analysis". Dr Luo has been recognized for his significant contribution to agricultural science and greenhouse gas research at the 2015 Hamilton Science Excellence awards.

Professor *Yan Li* has a world-leading track record in agricultural non-point pollution research. She has a background in environmental science with a focus on use of water and nutrients particularly in relation to upland agriculture. Professor Li is an experienced research programme leader and the deputy Research Director at the Institute of Agricultural Resources and Environment, Shandong Academy of Agricultural Sciences. She has led research projects for more than 20 years, including complex projects involving multiple national collaborators. She has been heavily involved in communication and extension of research results including being lead author of six books and more than 60 journal papers. Professor Li has received a number of national and provincial awards in recognition of her development of advanced technologies for pollution mitigation and excellence in agricultural research.

INDEX

A

B

C

D

E

F

G

H

I

L

M

N

U

V

W